Humanology

By
Michael K Chapman

A compilation of light observations on our modern way of life.

Humanology

Copyright© Michael K Chapman 2012

The right of Michael K Chapman to be identified as the Author of the Work has been asserted by him in accordance with the Copyright, Designs and Patents Act 1988.

All Rights reserved. No part of this publication may be reproduced, stored in a retrieval system, or transmitted in any form or by any means, electronic, photocopying, recording or otherwise, without the prior permission of the publisher.

This book is dedicated to my long suffering family who have endured long hours of reading, editing and spell checking on my behalf.

Preface

The substance of this book aspires to be an anthology of observations, idle conjecture and suppositions concerning the existence and behaviour exhibited by a very modern animal, the Human Being. Subjects are not discussed in great depth, but instead in a light and sometimes humorous attempt to add contemplation and comprehension to the behaviour exhibited or enacted by all of us on a day to day basis. Some of the antics, activities and habits are well known, but knowledge of others may depend on ones intelligence, awareness or apathy.

It is hoped the contents within may be described as a cocktail of sociology and psychology with just a dash of philosophy added for taste. Alas it may also be described as the incoherent ramblings of an opinionated individual. However, in order for one to form a judgment on the essence of this book, one first has to read it!

Enjoy.

Table of Contents

Introduction 7

Change 12

Cars and other motor vehicles 34

Shopping 52

Fashion 75

A Quickie on Attraction 87

Class and Breeding 107

They 119

Music 127

Television 147

Vices we love and hate 164

And so to summarise 185

Chapter One: Introduction

Firstly let me admit that Humanology is not actually a recognised word, yet! It has been fabricated in a vain attempt to make sense of the behaviour of modern humanity and the strange behaviour patterns that have arisen over the recent decades. The human being is undoubtedly a complex creature, emotional, vain, obsessed, frightened, arrogant, beautiful and ugly. There are those who value individualism and those that follow the herd. Those who choose to strive ever forward towards a full and adventurous life, and those who believe television is the answer to the mighty question: What is the meaning of life?

The human civilisation is a constantly changing ripple upon the surface of our planet, spreading like a tidal wave of insane ants. With each passing generation new directions appear, some can be easily explained while other directions simply provide a headache to the truth-seeker. In an explanation of the term Humanology and the fact that it simply does not exist, this book will endeavour to examine some of the more recent traits followed by the human animal. The intention is not to delve into the realms of deep philosophy or sociology or even psychology but briefly skim the surface of these multifaceted pools of human behaviour and examine an assortment of modern behaviour and trends in a light and sometimes humorous perspective. Behaviour patterns frequently over looked by the experts and theorists and bored academics. Our modern activities resulting from change, from modern shopping; from our sometimes strange fashion choices and even a quick peek on what we may consider attractive in each other. These are mainly superficial subjects but each one plays its part in our existence, more so perhaps than the normal ramblings of those experts who insist on discussing religion, politics and death as a method of entertaining us. Of course for all those with an interest or sadistic bent who may wish to discover more about the cheerful subjects of psychology, sociology and possibly even a touch of philosophy, the items discussed here may prove a first step into these unfathomable dark areas of conscious and unconscious cognitive processes concerning we wonderful, inept, confusing and unpredictable members of the human race.

And so to begin. I am writing this during one of my frequent bouts of insomnia, not that I suffer from insomnia; it is simply that once or twice a month I have trouble sleeping. I do not even mind these sleepless nights but I do worry about those around me who are sleeping and pray I do not disturb their peaceful slumber. It is very strange lying awake while all those near traverse the land of Nod and the house is quiet. I say quiet in politeness so I will not mention the grunts, snores, chuckles and farts that often disturb the silence of sleeping households. Nor will I mention the young men with car speakers blaring as they drive demurely (*remember I am being polite*), past my window in the early hours of the morning. Nor those people returning home late in a less than sober state whose whispered conversations and snatches of song issue forth at a decibel level rivalling that of a Town Crier. Peace during the night time hours is fast becoming rare, especially on Friday and Saturday nights. However at four in the morning, (*wasn't there a song about that?*)

one does expect some level of silence, with only the clicking and whirling of thoughts rattling round within one's head to disturb the quiet.

A thought occurs to me amidst my sleepless ramblings. Is it just me or is there no such thing as complete silence? Even when I lie still in the darkness of my bedroom between the sound of sleeping and wind movements, I still seem to *hear*, no actual physical sound can I identify but there is a kind of sound. I am not sure how to explain what I hear, a soft rushing, a constant whine, a very hushed whisper or whistle? I do not know, but there is always something there that is not true silence. Possibly it is merely in my mind as I am so used to noise during the day that perchance I am still hearing latent reverberation or echoes. Perhaps it is just my hearing, maybe I have an ear problem but it is not new to me, I have heard these sounds ever since I can remember. You might think that in the absence of any external audio stimulation, you would experience hearing nothing, yet with no apparent sound source around you still do. Many of us think of silence as being no external sounds, however sound can come from various sources. Air movement round ones ears for instance; in fact your whole body makes noises (muscles, heart, blood-flow, and organ function). Try contracting your jaw tightly while in a quiet place, you can hear your own muscles working. Over time our bodies change, as a child you may not notice these noises even though your hearing is good, before it degenerates with age. I certainly know my body has developed many strange sounds, such as creaks, clicks and grumbles as I have gotten older, and it is not all down to late night curries!

I value such nights, not for the lack of sleep but for the opportunity my mind has to wander, sometimes analysing past memories or events, sometimes simply dreaming of what may come, and sometimes pondering on many of life's little quirks or unexplained behaviour. Often my brain works like a private cinema screen replaying events of the day. I ponder things I have said but should not have. Things I had done but should not have, or things I have not done but should have, and of course those words I should have said but did not. As the saying goes, hindsight is a wonderful thing. Hindsight is also far more reliable than foresight!

But while I sip a hot chocolate drink in my little spare room or *aka* study at 4am on a Tuesday morning, my mind roams over vast acres of thoughts, some deep and full of meaning but not many, others like the chocolate drink for instance, are totally insignificant in the wider scheme of things. This night, without warning I suddenly began pondering the changes in life, changes that can surprise us when realisation hits home. Changes like noticing the absence of complete silence in the depth of the night. Changes that can cause sorrow, like the sagging bits of our bodies that have begun their travels south. I wonder if those living in the southern hemisphere notice floppy body parts heading north. It is just a random thought, if gravity pulled down to the south right across the globe, those in the southern hemisphere would soon find their belly buttons on a level with their noses!

In this attempt to justify the term; Humanology, it is hoped that the subjects discussed may reflect what others see in their immediate environment and in the wider aspect of the world as a whole. There will obviously be conclusions and assumptions that many will disagree with, this is guaranteed. There may even be comments or suppositions that many agree with, possibly the contents may cause some to shout incoherently before throwing this book from the nearest window! Again this is not only expected, it is hoped for. One can never listen to the opinions of another without reaching ones own conclusions of agreement or disagreement. Like the enjoyment of a hot chocolate drink, many will agree with its taste, many will not. Similarly some will consider the question of true silence to be inaccurate, others may consider otherwise. So if any of the material included in the contents offends you, remember we each have our own opinions but no matter how strongly we consider them to be true, others will not agree. It is anticipated many will find some of the arguments rest easy within their own perceptions of modern human behaviour. In the event of no one discovering suitable subject matter from these ramblings, one at least hopes the small humour may entertain. However the undoubted purpose of this literary concoction is to spark an initial awareness of how we as a race are progressing forward in a modern world.

Chapter Two: Change.

There are many changes in our world as we grow older, and I have just this moment noticed a change in me. I am drinking a chocolate drink! Strangely, I have always hated chocolate flavoured things, chocolate cake or biscuits, chocolate cereals and especially chocolate drinks! Chocolate itself I could eat by the truck load. I am a self confessed chocoholic, but never chocolate flavoured foods or drink. My acceptance of this rather nice hot chocolate drink, made by one of the larger chocolate companies signifies to me a change or a mellowing over the past years, an acceptance of previously hated foods, like sprouts. As a child I was not too keen on sprouts, much like most children, but again I seem to have mellowed even towards this humble and hated vegetable.

Many tastes change over the years, you can begin to dislike something you have always liked, or you can grow to like something you have always detested. Rationalising this statement, I think our tastes do not change perceptively, it's because as we mature, so does our understanding of what our bodies need for nourishment. Therefore we condition ourselves to eat foods like sprouts, ignoring that fact that we may still dislike them. Or it could fall into the category of an acquired taste, like beer. Many of us will remember the first sip of beer with distaste, few will honestly admit to liking the experience in the first instance. Pressures from today's social cast convince us that beer is a mark of adulthood and so we strive to enjoy the beverage simply to be acceptable. In most cases the acquisition of taste for beer develops quickly and stays for life. This I am sure is quite normal and experienced by many people, and judging from the present fashion of *binge-drinking*, I would almost hesitate to say beer and other assorted types of alcohol have become an obsession, and has little to do with taste buds which have long since been pickled.

Other things seem to change as we get older, not just those parts of our body succumbing to gravity but how we view the world around us. A fact verified by simply listening to the conversations of people passing in the street or work place. I have found it is wiser to just listen as they pass, following them and listening to their conversation as they walk will often result in some form of attack on your person. Memories play a huge part in how we consider the world has changed. We all have our favourite memory from the past, but some are more inclined to tell anyone within hearing distance than others. I overheard a young lad reminiscing about things he had done when he was just a little baby, not the grown up four year old boy that he was then. At the other end of the age scale, the elderly or more mature persons are famed for constantly using the opening statement of, *'In my day . . .'* or *'when I was a lad . . .'*

We have all heard the stories of how things were much better in the old days, people always claim they were happier then, and the sun seemed to shine hotter back then. But were things really better in the past I ask myself, what about all the progress we have made in technology, medicine and travel. Such progress allows us to watch even more rubbish on the television but now in high definition, travel to our intended destination faster and in comfort, providing we ensure a bank loan to afford the fuel! The advances in medicine have been astounding, it is a pity that due to government cut backs there are no longer enough nurses! But the fact is just how much we as a race have changed and progressed. The bygone days of peace and sunshine are in many cases simply glossed over memories, we embellish the memories that were happy and try to conceal those we regret or make us unhappy. In today's modern world, and I am talking mainly about the western world, if we need a doctor we can see one, eventually! If we need food we buy it, or even `acquire' it, from those huge conglomerates known as supermarkets that have sprung up from the ground like rabid gophers on holiday. There is no need to hunt our food or gather berries any more, no need to brave wild animals or wilder weather allegedly resulting from global warming caused by cows farting and possibly the odd car exhaust. Moving forward in time, children no longer sweep chimneys, few people now live in small cramped hovels. I take that back. Have you seen how small and cramped new build houses in Britain are these days? Even a rabbit would have to become celibate to live in one!

How we communicate has changed radically over the past decade or three. Electronic messages, emails and other unwanted communications fly the world in seconds and education is available for any who wish to study and force fed to those who do not. Even as near as the sixties, things we now take for granted were more difficult if not impossible than today. With the advances in medicine, technology and travel, our modern life is comparatively easy. I believe the dissatisfaction that many of we more mature persons feel, stems not from all the advances and changes that have happened and will keep on happening. No I think it's the fact that due to modern technology the world has actually speeded up. Not literally of course, or we would fall off, no it's just our way of living. Transport is faster, communication is faster, fast food, fast living, fast women (*I wish!*) The need to compete or just run parallel drives us all to strive harder, work longer hours, make more profit and generally do better than our peers.

There is much more pressure both at work and in the home to achieve as much as possible in a shorter period of time and for less effort. The need for more and more time saving technology has in reverse, restricted the time we have to spare. Seldom do you find a family where the partner goes to work each day, knowing he/she alone can earn enough to pay the bills and buy all the things modern living expects of him/her. Rarely do you find a partner who can afford to stay at home and look after children and housework secure in the knowledge that one income will suffice. No, in these days of fast occurring changes, habitually both parents work, rushing home in the evenings to care for children, do chores, prepare for the next day and finally head to bed exhausted with little time or thought to romance, quality time with the family or to enjoy relaxation in each others company.

Children's toys and hobbies have changed dramatically, gone are the wooden toys of old, the simple jigsaw puzzle, drawing pencils and Dinky or Corgi cars. Kids of all ages now demand the very newest technology. Mobile (cell) phones, personal computers and game machines are each desperately sought. The adults simply wanting peace and quiet rush out to buy the latest product to avoid those shouting matches and sulks from teenagers who really desperately need the latest fashion of sport shoe or trainers in order to maintain their standards and to avoid being shunned amongst friends. People in the western world and in the fast developing counties like China, India and Britain all search for the enlightened path, even if only to the television remote in their reclining chair. Even the very young children will cast aside the toy that is not electronic or multi-gadgeted, mortgage inflating and constantly media advertised.

So it is not that we were better off in past times, it was just a different criterion of living that has now changed. Today the pace of living is faster but we choose to live this way in order to achieve the standard of existence we require and to obtain the items we desire. Surely it was not a good life with bombs raining down on our major cities, food shortages, lack of medical experience or technology, communication slow and unreliable, rationing and depression and unreliable slow transport. Surly we would miss having no pollution controls and fumes filling the air night and day? There are many, many more items I could list here but I am sure you get the idea. All of us have memories we cherish, of times when life was good to us and living seemed easy. Just remember change is all round us and when our children are older, they in turn will reminisce about these very times that we complain about now.

I am not saying that the past was better or worse, life itself changes, and we just have to put things in perspective. As a child I would enjoy listening to older people recant their tales from the war years, from trips to other countries or just events from their past. No wide range of television programmes back then either, remember? A young person could become bored very quickly, even before the rise of the telly, so an old person recanting tales of their past was almost as good as listening to a radio but without the static and crackle of the pioneering radio receivers. One soon became accustomed to the farts, clicking of false teeth and grunts that normally accompanied such story telling!

As a teenager I was perhaps more disinclined to take time and listen to those more experienced in life than me, a surprise I know, a teenager who did not listen. What a shock! Sadly I was no different from many other young people in the past or the present because it is just the way of nature. All through the ages young people have strived to be different or rebel against the wishes of their elders, I was no different.

Time however, has a habit of catching you up, like developing a liking for the hot chocolate drink so everyone changes. As I mature disgracefully I deliberately attempt to stop myself from saying those immortal words:

'*In my day . . .*' or, '*When I was a lad . . .*' when talking to the younger generation. I do not want to be sniggered at, totally ignored or viewed with those pitying superior eyes that only the young seem to truly posses. Instead I have found myself working round the problem, I tend to begin sentences with statements such as, '*Years ago people would . . .*' or '*I knew someone who used to . . .*' Any way I can find to relate my past experiences to a younger person with out insinuating I am getting old! Such vanity!

The attitude of young people towards their elders has changed too. The automatic respect given by the young to the old and ancient has long gone. Most young people these days look upon the older generations as boring, wrinkly and a strain on the pension system.

Luckily this attitude does not apply to all young persons. It is mainly those who insist on hiding their faces from the world under hoods, or hoodies as they are more commonly known. Everyone has seen these strange beasts, bright warm sunshine, all the considered normal people in shorts and T-shirts and in amongst them all is the youth wrapped up in a thick hoodie and baseball cap, clinging to his/her idea of youthful fashion while sweating like a pig!

Through my job as a lecturer and my interests and abilities in music, I do mix with many young people, and therefore I find it easier possibly than most adults to associate with the younger generation. In fact I used to be one! (*Sorry, old joke*). On the whole I find the more intelligent, or perhaps I should say the more willing to learn amongst the young people are a pleasurable company. I changed intelligent to willing to learn because there is a difference. Sometimes *intelligence* is not worth the paper the qualifications are written on. I much prefer to meet and talk to someone who is willing to listen, to respond with an opinion of their own or watch and learn and contribute with those who may have something to give. Sadly in some cases, intelligence can also bring arrogance.

Not all the great minds in history are a product of University, or intensive schooling, in fact most are not. Learning can be from any source, but we are slowly *taught* to learn exclusively with our ears, teachers mumbling through uninteresting lessons, lectures and speeches, and we develop a language based form of communication, despite the fact that a high percentage of communication is visual - body language. A nod, or wink and especially a smile can speak volumes to the person in receipt of such a gesture, gestures also save a multitude of words, one, or if generous, two fingers can make plain what the gesturer is thinking. We forget how much we use our body when talking or communicating, some people rely heavily on their hands when talking. Tony Blair for instance, our erstwhile British Prime Minister. One must ponder if he would still be able to speak if his hands were tied behind his back, but let's face it, if his hands were tied, someone would take the opportunity to gag him as well!

As a child, communication changes rapidly. Initially adults tend to speak absolute rubbish to very young children. Meaningless words such as *googoo gaga* appear to be the normal language of the infant. This then changes to words like *Choochoo train* and *horsesworsy*. Confusion strikes the child later when suddenly the adult is admonishing them for the use of such vocabulary, demanding they talk correctly by using the words, train and horse. Just when the child has got over the fact that it had to learn by audio means, we throw it back into confusion, by introducing change in the form of a written (*visual*) code, which also has to be studied.

Having the ability to learn is far more important than remembering answers to questions in tests and exams. I have always considered the fact that a person's ability to achieve good results in a question and answer test or examination depended more on that persons memory abilities than actually intelligence. The capacity to learn and the willingness to learn will often succeed where academia fails. Learning is a lifetime occupation, we continue to learn through out our existence as we adapt to changes in our lives and those around us. Please do not confuse intelligence with wisdom either. A car mechanic often has a much deeper understanding of the vehicle he is working on than the so termed intellectual who designed the now broken vehicle in the first place. A long term patient can have a fuller understanding of his illness than his doctor. A fact increasing noticeable in today's world, where information and diagnosis can be found at the touch of a button. Intelligence is often a label for those with academic standing, but wisdom comes with experience, of adapting to change, learning is life long and something all of us do and should continue. Of course most of us know that the highest form of intelligence is common sense!

Young people, as we all used to be, do not consider change as an important factor in their lives. It is not much use telling a sixteen year boy that he should consider his pension! As far as he's concerned, old age is a millennium away. Telling a child he or she may eventually like the taste of sprouts will most likely be met with sincere disbelief, and maybe the odd howl of hysterical laughter! Children and young people in the twenty first centaury accept change in subjects that relate to them, modern technology for instance. The young have no fear of smart phones, computers, game machines, DVD's or other such technological wizardry. Those of us more senile old fossils often find ourselves seeking out the assistance of a younger person when dealing with technology, like the seven year old lad helping his parents understand a mobile phone. Long term change affects them differently, planning for change when leaving school, planning for change in lifestyle when setting up home on their own, or accepting that their bodies will change and life insurance will become important is not what they deem as a necessary factor of consequence in their present moment of time.

As I continue to ponder while writing and sipping my now lukewarm chocolate drink, I think back on the changes I noticed during my youth, my school days in particular. As a child I was often a `sickly child', rarely well enough to participate in the normal activities of childhood. While my friends became tanned and healthy playing football and running around like rabid dogs, I was restricted to drawing, which I now hate, making plastic models or reading. Looking back I think it was then that my mind first began to wander widely. Or possibly the initial onslaught of insanity! As I watched the other children from a window, the sun beating down and the whole world seeming to shimmer in a wonderful warm yellow haze, it was my mind that joined those cheerful children playing outside. I know, I have just reminisced about how good the old days seemed, but that is how I still remember them. I played all the normal childhood games, football, rounder's, kiss chase, cowboys and Indians and soldiers, and of course I always won the game, won the girl and won the hearts of my fellows in these games, but frequently only in my head! I did manage to join in when ever I was fit enough but sadly the actual results of my football skills and attempts to win over the hearts of the girls failed miserably in real life. No wonder I preferred my dreams. I was either a very sad case, or a nutter. I know which one I would choose!

I eventually discovered a talent that put me in high standing with my fellow students, due to the amount I read I quickly became something of a story teller. I regaled my friends with stories such as *The Devil Rides Out* by Dennis Wheatley plus items by Edgar Allan Poe, embellishing the scarier parts depending on the reaction of the prettiest girl listening, attempting to show how brave and manly I was. It never worked though. . . . *Quoth the Raven "Nevermore." (Edgar Allan Poe – The Raven.)*

Please do not assume I am looking for the sympathy vote here, I mentioned it only as a phase in time when I began to think more intensely about the world around me and the people who inhabit it. I am not trained or qualified in matters of the mind, nor am I a experienced sociologist, so please do not expect lots of scholastic wisdom within these pages, I am only stating my interest in things I observe round me and the thoughts, sometimes bizarre I'll admit, that go with them. I did mention I am not a sociologist or psychologist; in fact it's pretty hard to think sideways when one has a background in science. But one cannot discuss thoughts and speculations of the mind or behaviour without delving into some level of theoretical supposition.

It was during my school years that I developed an interest in watching people and trying to anticipate what changes their world would bring. Watching the sports mad teenager who had little interest in learning or building his mind, he only wanted to develop his body, enjoy physical activity and have fun. Nothing wrong with this I hear you cry, and no there is nothing wrong with this, it was how he wanted to live his life. He was a nice lad and could achieve amazing feats of balance and dexterity. He once attached roller skates to the bottom of a pair of wooden stilts and could use them to negotiate steps and awkward areas with ease. He played football and cricket like a professional, in fact watching our present national teams he could do a lot better. His only draw back with all this activity was his extremely smelly feet! I did meet him again recently and was pleased to discover he had found his path in life, not a famous football star but a loving husband and father and more remarkably, in full time employment! So kudos to him!

Then there was the boy who considered himself a dashing young man, always combing his hair and preening his appearance. Again nothing wrong in that, he had simply chosen a different path and was more concerned with his appearance and attracting the interest of the girls in school. Certainly no arguments there as I could easily relate to his choice of destiny, in fact I was often to turn a shade of green with envy at his success! I believe he is now quite big in the insurance industry these days and is highly successful. And I say kudos to him also. It seems to me that his consideration to his appearance may have played a big part in this success, following the changes he made as a teenager. Another conduit chosen by a lad at school was that of learning, he thrived on knowledge and left us in his wake academically, managing to understand often the most difficult subject with apparent ease. Not too good looking, not interested in sport or his looks, he still managed to obtain a female following amongst the girls with similar interests in knowledge.

As for myself, I did what I could. I quickly became known as a *shoulder to cry on* as well as a raconteur. The other kids would often come to me with their problems, seeking answers I was too young and inexperienced to supply, but I tried anyway. I gave comfort where I could but mostly I just listened. Often sharing a problem or discussing it with someone helps the sufferer to better appreciate their own situation, and choose the correct path to take them. Least that is what we are told by the psychologist if you are bullied or melancholic. We are advised always speak to someone, though I would suggest you do not talk to the bully! I like to think I did help in my own small way, certainly many of my school friends continued to bring their problems to me so I must have had some triumph. Plus I did manage to get my arm round the shoulders of some very pretty and otherwise unreachable girls as they sobbed out their concerns on my permanently wet shoulder

Changes affecting our growing youth, our taste buds and our view on the world as a whole are frequently pushed to the rear of our minds, the pressures of modern day to day living forces us to focus on the present rather than dwell on the past. But there is one form of change that is guaranteed to bring forth streams of verbal displeasure and confusion. This change is simple but devious and has a proven track record in parting us from our hard earned money. There is nothing sinister in this form of change, however almost all dislike it, moan about it and frequently find it totally infuriating. When this change occurs it throws us all into hysterical confusion resulting in aimless wandering, forgetfulness and an utter sense of helplessness. This vicious attack befuddles our senses, destroys our confidence and deflates our natural assumption that all is well with the world. OK, so what is this change? This change sneaks up on us in the most simplest but devastating of forms.

As we head off to undertake a shopping chore, we meander unsuspecting into the same shop or store where we are regular customers, and suddenly the full horror hits us. All the food on the shelves, the washing products, cleaning equipment, pet food and even the toilets rolls have changed position! The store has changed its entire stock range to different shelves and locations within the palace of groceries and essentials; nothing is in its usual position! Suddenly no one can find the items they seek, the bread is where the eggs were supposed to be, luxury chocolates replaced the toilet rolls, tins of beans miraculously became tins of peaches and tubes of haemorrhoid cream lay ready for the unsuspecting shopper as they reach for the toothpaste that is no longer there! Everything in our once favourite store has been changed, maybe by small malicious Pixies overnight or someone with a dread sense of humour.

However the idea of changing the stock layout is simple, people get used to where all their favourite items are like spouts and toilet rolls. When entering their usual venue of grocery purchase, most shoppers can be seen heading directly for known items in familiar positions on their shopping list, but ignoring the myriad of offers waiting to temp the money from their pockets. This behaviour does little for the small store keeper, multi-conglomerate super market share holders or even the store manager's prestige as we all walk blindly past the latest offers and new lines of merchandise. So the devious plan to rearrange the stock is intended to bring buyers notice to other products within the store. It is an attempt to increase our shopping selection and of course, spend more money. This is especially relevant to the stores special offers, how can they be a special offer if no one notices them. It has been known for an innocent shopper to wander blindly into a favourite store, only to emerge an hour later with not one item on the original shopping list.

Instead the handfuls of plastic shopping bags contain goods never previously considered. Bags filled with money off bargains, buy one get one free, half price and other such customer incentives, and so the bewildered shopper heads home.

Not until safely within the familiar surroundings of the family residence will the poor customer realise what is actually in their shopping bag. Not until the partner demands to know where the items on the list are, the kids demand to know what happened to their sweets. Grannies enquire on their missing pipe tobacco - grandfathers complain on the absence of knitting wool! In this twenty first century, despite all the marvellous changes that have occurred, in technology, in medicine, in transport and in ourselves, this is one change that affects us all, infuriates us all, causes confusion and sends customers running screaming from the shop, pulling at their hair in frustrated handfuls after searching in vain for the item they required.

Changes affecting our bodies comes somewhere under those annoying things. How we change over the years, our attitude, our situation and of course our body. The two greatest changes involve weight and puberty. Once, long ago in the dim distant past, I was a slim young man, what the hell happened! I now look like I ate the slim young man with a side salad. My stomach is insisting on meeting my knees for lunch, my skin is becoming baggy, wrinkles, aches, pains and other strange inflictions attack my body on a daily basis. Standing up from the sitting position is now accompanied by audio creaks from limbs and verbal outbursts as stiffness reduces the speed of movement.

I realise this is a change that every living thing must go through, hopefully. Even a blade of grass would wish to live long enough to accomplish old age, and we humans are no different. The onslaught of age and the changes that occur during this process are plain for all to see. Attitudes change along with age, the elderly have no embarrassment about breaking wind in company, speaking their mind to absolute strangers, complaining about everything and caring not a jot about the thoughts of others. With age comes great wisdom, or senior arrogance and cantankerousness. Each of us strives for wisdom, few succeed. Changes in ones perception of life can prove interesting, like the old man fast asleep in the corner of a bar with false teeth hanging loosely from his mouth and trousers gapping between brightly coloured braces, or gentleman's suspenders, which ever you like to call them. His outlook towards how other people see him has long ago changed into *I do not give a damn* phase. So many things change as we get older, our bellies expand exponentially and begin the travel towards our knees, our chins multiply and sink towards our chest, our desires change from pretty young girls to thermal socks. Our expectation of what a good day constitutes changes from a successful day at school or the office, to finding an unsuspecting victim to accost with our memories of the war, how much change one would get from a shilling and how the world was a better place when they were young.

Life causes big changes in our lives as we get older, it also creates huge changes in our bodies and how we perceive the modern world. But obviously it is not only age that changes us as individuals, changes also affect the young, especially the dreaded puberty.

I remember vaguely when the changes brought on by puberty attacked me, dismissing the small innocent boy and admitting in his place, the terrible teens! I was a teenager, the most feared of all life's stages. The delightful puberty changes us in ways we simply do not recognise at that time, we probably would not listen anyway. By the time we reached the teenager decade, we already knew everything and would never confess to being unsettled by the presence of girls blooming into womanhood.

With the coming of my older age wisdom, I now feel I learnt much from observing others during my school years. I watched and wondered about the interactions between members of staff in the school. Too young and innocent to know what was happening, I played items I had seen over in my mind as I attempted to make sense of them. A flutter of eyelids, a shy glance and even a slight blush between male and female staff members. Certainly I now understand the art of flirting, but back then my experience was limited to say the least. Unquestionably I understood the juvenile flirting practised by those initially discovering the meaning of the word randy, as our bodies changed from innocence into puberty. But adult flirting, especially among the staff was by necessity more understated, they considered their actions to be un-noticed by the pupils, but it was!

Many uncomplicated and amusing actions by my fellow school friends caught my eye, young girls and boys falling out because one had mischievously or not, as the case may be, punched or hit the other. Again with experience one realises that children can have feelings of attraction or crushes towards each other but not know how to cope with them. They wish to touch or make contact with the person they are attracted to but have no understanding of how to go about it. This would then lead to one hitting or playing a prank on the other in order to attract their attention or purely as an excuse to have physical contact. Unfortunately a quarrel or squabble would result in resentment when all they really needed was a hug or to develop a stronger friendship. Times really became interesting when the initial puberty changed to full sexual maturity!

As adults we all understand what it's like to have a crush on someone, least that's what we call an attraction between two young people. I remember having a serious crush on a girl at school, a very painful experience at the time as I had no knowledge of how to deal with it or even what it really was. My dreams were filled with images of this girl and my waking hours were spent trying to get as close to her as possible without portraying my interest. Later I realised why I had a crush on this girl, she was not too good looking nor did she have a particularly good figure. However she was constantly nasty to me and in my efforts to make her like me more, I developed the crush. I soon got over it!

A youthful crush is quite normal and mostly no harm comes from it. Too often though, when a child discloses the big secret to an adult, the reaction is sceptical to say the least. Adults quickly forget what it is like to be a child; J. M. Barrie the author of Peter Pan realized this and turned it into a book. To a child or worse, a teenager, a crush is a terrible thing when not reciprocated. Adults call it love when the very same feelings for another person befall them. Unfortunately either as a child or an adult, you cannot choose who will become the object of your affections, often there seems no logical reason why you are attracted to this person. It may be that you have lots of interests in common, it may stem from spending long periods of time in their company, or it could even be their personality. They may make you laugh, they may be kind and sympathetic, like a nurse for instance. It may be a slight hint of fear toward the other, like hostages falling for their captors, mostly though it's that they are pretty or have a great body! Today's children have a better perception of physical attraction between male and female due largely to the easy access to information on TV and the Internet. I am not sure whether this is a good thing or not, so I will avoid this subject for now.

As a child I still remember the odd actions and expressions I saw when observing grown ups together. Observably it was easy to tell when an adult was angry or upset because it was normally over some thing I had done! I know I said earlier that I was a sickly child, but not constantly. I will admit to being a cause of frequent headaches for my boarding school headmaster when I was healthy. I was just one of those children who always managed to get caught up to something inappropriate. It appeared unjustifiable; many of my friends could fall into a pile of manure and still come up smelling sweet but not me. Those expressions I understood but when a pretty young member of staff came into the room, things changed, especially among the older boys and any male members of staff. At the time I had no idea, but obviously I do now.

Most of us can remember situations such as this. When friends or relatives came to your home for a party or gathering, perhaps after a wedding or celebration, and then the wine begins to flow. Christmas

Eve, New Years Eve or birthdays also lead to such gatherings. At first all would be calm and conversation would be civilised and deferential. As the evening worn on, a change would take place. Talk would become louder, laughter less forced and polite, the music volume would be turned up. Eventually everyone become more relaxed and free with conversation and their conduct! All too soon the children were bundled off to bed, after suffering wet and peculiar tasting kisses from mothers, aunties and grannies, plus the odd person they had never seen before and who turned out to be a work colleague or distant relative. Sleep of course would be impossible as the noise and laughter grew in the rooms downstairs. Sounds of clinking glasses and opening cans along with frequent trips past the bedroom door as the adults need for the toilet increased. Sometimes one would hear muffled whispers as a couple passed the door, obviously tired and looking for a bed to sleep in, or that is what young children conclude.

The final surprise would come in the morning, when children would watch and wonder what must have happened the night before that changed their parents into bad tempered grizzly bears, bears with a headache and requiring frequent and some times frantic trips to the over worked toilet. Frequently being a child observing parents and adult behaviour could be extremely confusing!

Change is inevitable in our lives and in the world around us. We all change during our lives as consciousness and understanding of the world and its inhabitants grow along with time. Like the hot chocolate drink or the poor sprout, our tastes change. Our understanding of adult behaviour begins to make sense as we eventually become adults ourselves. These thoughts of change have travelled though my mind and now my cup of hot chocolate lies cold and empty upon my desk. I have made an insignificant attempt to describe how change affects us all while sharing some of the thoughts, impressions and observations I make as I wander through my mostly uneventful life. Many experts have written books on thought and the inner workings of the human mind. I do not intend to attempt such a task, I am only chatting to the reader about my perspective on life and how we live it as individuals.

So I have no excuse nor do I give one for my enjoyment of this hot chocolate drink in the early hours of the morning. I have changed in ways, certainly in looks, in my style of clothes and in what I drink. Be prepared, it happens to us all!

While I am on the subject of change, I wonder how we now view the infamous motor vehicle these days. The automobile has certainly changed over the last century or so and alongside it our attitudes towards this mechanised tin can on wheels has also changed. I certainly now expect certain comforts in my car, such as electric windows, a good heater and of course an excellent radio and CD player. A complete change from my first adventures in a motor vehicle with a small transistor radio perched on the passenger seat beside me as I drove. Driving without some thing to shout at is, in my opinion, totally unacceptable! Part of the pleasure of driving is being able to disagree, argue and shout comments at the lack of intelligence demonstrated by radio DJ's, chat show hosts and the opinionated plonker's who arrogantly give their opinions over the air. Our whole world has changed largely thanks to the industrial age and the motor vehicle. From automobiles, to trains, from motorcycles to planes, the world has changed into a much smaller place.

Chapter Three: Cars and other motor vehicles.

OK, so let's have a chat about cars. Earlier I spoke of how I like to sit in my car and watch the world and its people go by, watch and form opinions, right or wrong about those who pass by my wide screen viewer. Along with shouting at my car radio of course! Cars or automobiles are a set standard feature in most countries of the world now, China and India fast overtaking the western world in the numbers of cars owned by ordinary working class people. Well maybe not quite as many as the USA yet. Russia is another country coming up in the fast lane of car ownership, no longer are the only cars in this country seen to be long, black and sinister limos with darkened windows and carrying party members or officials. Wealth, both personal and economical is growing in Russia, allowing even the commonest comrade to own a car. Mass produced motor vehicles first rolled off the line by Henry Ford in the USA, designed as a cheap, safe and reliable form of transport for the ordinary people, enabling them to travel freely to work, home or vacation. I have a Ford at present and I swear that it must have been made by Henry himself, the trouble it is giving me!

At first a motor vehicle was seen as a tool, along with a good tractor, reliable economy and healthy family. Motorcars replaced horse drawn carts and wagons. Lorries or trucks began to share the workloads with the railways, even small motorcycles had their niche in the growing demand for mechanical transport. Roads grew in number and use with drivers expecting a smooth wide surface, a ribbon of black that would give access to any part of the country.

Obviously potholes had not yet begun to breed when these expectations were high. No thought was given at that time to possible world pollution, in the early days of the motorcar; pollution was probably an unknown word. These days it appears to be heard by all but ignored by many. So the horseless wagon, motor car or automobile sprang into life with a splutter of smoke, a whiff of fumes and road rage followed with a curse! Motor vehicles of all shapes and sizes, including the Sinclair C5 (*almost*), have been churned out in ever increasing numbers to satisfy the need for travel, speed and transportation by most countries and populations ever since.

Motor vehicles have changed the world, they have made the world a smaller place, allowed access to parts rarely visited by civilisation, increased the pace of life and enabled a wider distribution of goods and wealth. Our cars, buses and trains have contributed towards the joining or mixing of many different races as visits or relocation to other countries became feasible. Paradoxically, as ownership of such vehicles increased the ability to travel freely has become slower and restricted over the years. Now the sheer numbers of vehicles clog our ribbons of road and have succeeded in putting a stop to the freedom of the empty highway. These days simply getting from A to B requires logistics, satnavs, tranquillisers and a lottery win to afford the fuel!

The initial thought behind all forms of mechanical transportation was to enable mankind to travel or transport goods much more freely. The motor vehicle was designed as a tool to help with the expectations of living, working and playing. I agree that dear old Henry Ford had the common man in mind when he first developed the assembly line, mass-producing cheap motorcars for the populace. With a fast reliable form of transport that replaced the poor old horse, man could achieve his tasks quicker and thus give himself and his family more time for pleasure. A tool to help mankind, a step up into the future, progress at its best, opening the door to the wide world and all its inhabitants and resources. Motor vehicles also very quickly became one of the world's most lethal weapons since the invention of the gun or the atom bomb! So, when did this simple tool, this movable lump of metal, plastic and rubber become an item of worship?

The more time I spend sitting in my own car watching others around me, the more I wonder what is this strange fascination and pride humans have with their motor vehicles. Vehicles and especially cars are held in extremely high regard in the mounting list of possessions owned by a single person. We do our best to protect them, to keep them clean and shiny, to foster their best interests through yearly servicing, weekly washing and polishing. We take trips to the local garage for even the slightest strange sound or knocking, though removing ones mother-in-law from the rear seat often reduces noise levels within the vehicle. Whole neighbours empty onto the streets, drives and garages every weekend as the DIY's dive happily under bonnets (hoods) and lay prone on the ground as they tinker with any broken part, rattle or squeak. We lavish money on them like no other possession we own, mistresses and toy boys included. We show them off to the world and worry about their well being when we are forced through necessity to leave them unguarded in a strange car park or parked on an unfamiliar road.

When parked our protection of the motorcar far exceeds any other factor including personal safety, the safety of others, the convenience of others or the simple provision of access for other road users and pedestrians. Women and children, blind people, old people, disabled people, people with large amounts of shopping, people in a hurry and delivery persons. Everyone who has a right to use the pavement are forced to walk round our prized possession as it sits gleaming half on and half off the pavement, its hazard flashers blinking happily to the world, thus seemingly justifying its right to be there.

Huge amounts of money is spent on car security, car alarms that attract no more attention than irritation to those in the vicinity, speaking alarms that warn passers-by to keep clear of the vehicle or the police will be notified, alarms that set lights flashing and horns blaring when a pigeon deems to land on the car's shiny roof. Our desperate attempts to protect this tin can on wheels often wanders into the realms of obsession, unfortunately this obsession does not extend to the early hours of the morning when a stray cat triggers the blaring car alarm. Then the owner decides sleep is more important than his beloved vehicle as he turns over in bed, and ignores the screeching sound of the alarm emitting from his car parked on a street lined with houses and now wide awake residents.

Modifications costing even more money than the entire bank funds of a small country are fitted to many vehicles in a vain attempt to personalise, or customise this highly worshiped collection of future scrap metal. Stereo systems boom out music to all in a mile radius, the owner attempting to proclaim his ownership of the vehicle and his individual identity or his commitment to a culture, cult or fashion by the style of music he plays. The owner of the music system is also attempting to attract members of the opposite sex, but only the deaf need apply. Flashy paint jobs or increased number of lights, stripes, wide wheels, chromed exhaust pipes, furry interiors and car stickers proudly display the owner's niche in this life. Onlookers generally have a different opinion of the person behind the fur covered wheel of the gleaming monstrosity, the vehicle proclaiming to all that the owner is an idiot with more money than taste!

Engine performance, tuning and added extras all help to increase the vehicle speed, young people especially seem to believe that their vehicle is the only one like it in the world, its uniqueness mirroring how they view themselves in today's society. Again it is often young people trying to attract the attention of others, mainly those of the opposite sex, or middle aged men in a vain attempt to appear young. Older drivers, specifically grey haired men with scarves, flying around in classic or vintage sports cars. They too are often only attempting to get noticed, again by those of the opposite sex, or the same sex as by now they are not too fussy. Most car users attempt to personalise their vehicles in some way, even if in an insignificant way such as a bumper sticker or rear window adornment. Farmers appear to collect as much mud as possible on their vehicles, to be deposited on the main roads when ever possible. Construction workers also follow this trend while travelling salesmen are happy to gather flies and other assorted insects on their windscreens and all mechanics have a tendency to adopt the greasy hand print design.

Then there are the beige people, those older gentlemen whose automotive possession gleams dazzling in the sun, blinding other road users and pedestrians alike with the high gloss polished paintwork. These drivers cause more fury than tractors and caravans on the roads, as they refuse to drive their status symbol over 30 miles per hour, no matter what the speed limitation of the road is. Heads turning neither left nor right, interior mirror used only by the wife for makeup purposes, the beige driver cares not a jot for other road users, he is king when in his car.

Trundling along the middle of the road next is the four wheel driver, huge monstrosities of modern motorcar initially designed for off road use by farmers, explorers, veterinary surgeons and the police and military forces. Sadly these vehicles quickly became fashion accessories for the rich, the 'wannabe' middle classes, the pompous and ridiculous. Mothers in cities decided that a massive 4x4 supertanker was the only way for their children to travel on the half mile trip to and from school. When questioned these mothers, not forgetting the fathers, all insist a large vehicle is necessary for the transportation of the offspring, a normal sized car would absolutely not do! I really feel sorry for my two children, both now grown, who had to suffer the cramped conditions of the standard family salon, with just four doors, a boot (trunk) and a medium sized engine.

School parking areas, road sides and especially in the No Parking zones near the school entrance will overflow with these motorized monstrosities twice a day during school term times. Yellow lines, No Parking signs, yellow zig zags are all ignored as the driver and its small occupants spill forth and rush through the school gates.

Traffic queues in every direction as the modern essential form of school transport blocks the street to all other users. When one rolling road block finally moves on its way to the nearest out of town supermarket, another quickly takes its place. The driver waving cheerfully to the immobile fellow road users and gives the traditional identification call preferred by this ignorant breed,

'Sorry. I'll only be a minute.'

Everyone waiting knows this is not the case as the minute stretches into ten minutes as the driver simply had to discuss a matter of vital importance with the first available unsuspecting member of staff that wandered within range.

The rugged four by four wheel drive vehicle can often be seen lurking on clean drives, wealthy avenues, outside restaurants and wine bars but never in fields or mud tracks. Built to withstand and traverse rough terrain, these vehicles are now polished almost to bare metal and no speck of dirt, no grain of sand or smear of mud is allowed on their gleaming surfaces.

A more acceptable form of school transport is the people carrier, a vehicle designed to hold seven or more seats while keeping the exterior proportions of a standard large car. When passing one of these vehicles, one often has to peer closely to glimpse a very small bundle of childhood securely strapped down and lost in the vastness of the interior. Again the owner will insist that a large vehicle is absolutely necessary when one has a small child, an extremely small child!

This pride in the possession of a motor vehicle is not confined to motorcars alone; motorcycle owners can be just as fanatical about their hog, chop, street racer, off road or scrambler, scooter, moped or custom machine. Brand new or classic, race performance or tractor, every form of motorcycle is worshiped at some stage in its existence. It must be said that motorcycle owners can be even more fanatical about their rides, I know, I am very much the same. A motorcycle is much easier to customise than a motorcar, after all it mainly consists of just an engine, frame and two wheels, and so much can be done to achieve that one off personal look. Television programmes have sprung up all over the TV channels showing custom cycles, each one proclaiming their machine is faster, more powerful, shinier and more unique than the last. The most popular motorcycle show at this moment in time appears to be *American Chopper*. A programme following an American family run business, churning out radical performance customised motorcycle machines for wealthy customers or simply for show. Strangely though, I think the popularity of this show relates more to the on screen battles and arguments filmed between members of this host family rather than the attraction of the motorcycles.

The make of motorcycle apparently sought after by every biker in the entire universe is of course the Harley Davidson, an American built motorcycle with a V-twin engine, the design of which has barely changed over the decades this model has been in production. Nearly all the televised motorcycle shows heap praise and glory on each and every Harley Davidson machine that rattles in front of the cameras.

Motorcycles seem to hold respected places in most people's memories, often more so than the first car and certainly in many cases, more so than the first girl friend! Sadly I am getting old so I barely remember anything these days, but even I still remember my first motorcar and my first motorcycle, not much else though, unfortunately . . .

In my middle to late teens, my brother and I went through the ownership of several motorcycles, now deemed to be classics. I can remember working in the sun at the back of our shed, trying hard to pretend I actually knew what I was doing as my brother and I fiddled, cleaned, tinkered and polished old clapped out machines that are now worth a small fortune with collectors. Elite motorcycle names such as a James, Villiers, Greaves and Velocette. Not forgetting BSA, Triumph and other such two wheeled mechanical treasures of the time.

Owning a British bike was paramount back in the sixties and early seventies, the onslaught of the Japanese machines were held in the highest contempt by all serious bikers, Rockers and Hell's Angel clubs. I was no exception, I had acquired a Honda 90cc motorcycle come moped by default which caused me no end of embarrassment and was quickly sold off for spare change at the earliest opportunity. Strange how things change, I have owned many Japanese machines since, I still have one now, and to be honest I cannot remember why I bothered so much with the old British bikes. Even being some 27 years old, my Honda still starts at the push of a button, has no oil leeks and rarely breaks down as soon as I get it onto a main road or motorway or it rains!

There are many, many brilliant motorcycles in production across the world today, high speed Japanese racers, Italian superbikes and even a resurgence in British high performance bikes like Triumph and Norton. Overall reliability has improved immensely, performance has evolved into an art, every day new makes and models are making waves in the world of motorcycling.

Anyway, it is time I got back to the main concern of this chapter, motorcars. Obviously I own a motorcar and I like to ensure that to the best of my abilities it runs reliably and looks clean, shiny and reasonably tidy. Unfortunately the latter is hardly ever the case, in fact I seldom clean or polish my vehicle, though I do my best to keep it mechanically sound, after all, what is the point in owning a mode of transport if it will not transport! I have a fair knowledge of mechanics which helps, plus I have some very gifted friends who are almost always happy to help, and I know of a couple of garages reliable and cheap enough to afford without needing a second mortgage. However that is as far as my own preoccupation with my motorcar extends, I can manage without one and have done so on many occasions in the past. I personally regard my car as a simple form of transport, a tool to make my life easier, and something to carry the weeks shopping in. This simple degree of attitude for a vehicle is unfortunately not held by everyone, for many their car is akin to a God.

So why do we hold our motor vehicles in such high regard? Why do we worry so much about the general appearance? Or the speed capabilities of our little mechanised transport? What matter if our car more expensive than our neighbours? Why do many of us strive to be seen owning the latest model? What is the reasoning or logic behind having the most gadgets? And finally, where in the hierarchy of our possessions do we rate our vehicle? Using an Internet site I asked this question and here are some of the more printable replies.

An answer that came up frequently was *freedom*, though this statement was never expanded on, freedom to do what exactly? Independence appeared often as well, again I ask, independence to do what? Freedom is a strange word often bantered about with no real regard for its actual meaning, often those who cry `freedom' the loudest are in fact already free! It is the quieter, secretive, plaintive cries for freedom that really mean the most, not the self opinionated shouts of those who have had no cause in their wealthy, civilised lives to use this word in earnest. Freedom is a word that brings to mind Nelson Mandela, a prisoner for twenty five years simply due to his political beliefs, freedom is a word thought but not uttered by residents of many countries of the world, countries still run by those who would physically and violently suppress others. Freedom is an ideal that many under privileged, starving, sick and those suffering extreme poverty can only dream about. Freedom is not a word I would use to describe a simple form of transport!

But owning a vehicle can give you the means to escape I hear some of you scream at this book. Wrong! Says I; owning a vehicle with no money to pay for fuel can be a slight hindrance, even the very fact that one may own a car but have no money for food makes that car less than useless. In the dictatorship ruled countries, if one did have the funds to acquire fuel, would they be allowed to use it?

Would they be allowed to leave their country, alive? And reaching down to the very bottom of deprived, degraded and desperate humanity, would they be able to maintain ownership of such a prized possession from even their own peers? No, a car is only a tool, it does not give freedom, not in the real sense of the word, nor does it give independence.

So what is independence? Why is it so important? What real independence does one obtain from owning a motor vehicle? As soon as you get into a car you have to follow rules, you are not acting independently anymore. Laws tell you where you can and cannot drive, laws dictate through safety obviously, what side of the road you drive on, what speed at which you should travel, who has the right of way at junctions, crossroads and roundabouts. This is repeatedly a cause of some confusion as many drivers simply do not know their left from right or arse from elbow! Many wealthy drivers expect an automatic right of way based on the value of their vehicle, and a certain number of non male drivers still expect a courtesy right of way based on their sexual gender and not the legal laws of the road.

Traffic lights tell you when to move and when to stop, borders and barriers tell you which country you may enter and those you cannot. Each individual circumstance limits the range of any small independence; have you enough fuel for instance? Can you afford any more? What do you do if your vehicle breaks down? Can you afford the repair bills or the replacement vehicle? Do you have the time and the resources to reach your chosen destination? Are the roads blocked with other traffic, all attempting to proclaim their own independence as they sit for hours in traffic jams going no where! So is the motor car really a form of independence or has it become more of a hindrance?

"I need a car to get to work," is another frequent claim, and yes I would agree that using a vehicle to expand one's range of employment opportunities is a very good reason for owning a car, but freedom? Or independence? No, in fact what these people are really stating is that they use another form of tool as part of their continued employment and that tool is a motor vehicle.

A further valid explanation for our continued reliance on personal vehicles is because they help us get to all the activities in our lives, making communication with family, friends and other acquaintances more accessible. Again this is a frequent statement and it has valid points. However one must ask, if it were not for the motor car, would we chose to live so separated by distance from our friends and loved ones? Would we deliberately undertake activities outside a suitable radius of obtainment? Has not the motor car actually made a niche for itself and caused the pace of our lives to speed up, become more complicated and certainly more expensive?

So does the ownership of a motor vehicle truly give us independence? In some small personal way I must agree it does, however when one considers the meaning, the true meaning of independence, can we justify using this terminology in relation to a simple tool? Freedom and independence are very big words to millions of people in our less than perfect world, can such an ordinary tool really be described as such?

I believe no man made transportation device can provide freedom or independence in their truest forms. The ability to travel via mechanical means over our landscape cannot seriously be associated alongside two of the most powerful words in the human language, freedom and independence.

Earlier I mentioned attitude in regards to personal ownership of a motor vehicle and I intend rambling about that now, after some good words about our cars for a change. OK, so what can be said in support of our love affair with personal motor transport? By love affair I do not mean sexual activities with an exhaust pipe. No I mean the care and attention we give to our vehicles. Anyway what can be seen as benefits to owning a fuel guzzling, fume producing, money eating, pedestrian threatening metal monster on wheels? Well one thing comes to mind, saving time! I have just come back from a trip to the vets with my little dog. The vet's surgery is some six miles away but it only took me about fifteen minutes to get there. How long would it have taken if one had to walk? How long via public transport? Assuming there was any available public transport within my window of time. More to the point, could my little dog make it that far? In the comfort and dry of my car, me and my dog made the trip with little effort and minor expense, the Vet took care of the financial shock. Saving time and effort is probably two of the greatest reasons for owning one's own motor vehicle.

The motor vehicle gives us the ability to move around, to visit, shop, work and travel without the added expense occurred by other transportation methods. Notice I did not say that the vehicle gives us freedom. I am very happy to admit my life is easier with the ownership of my personal transport, it gets me from A to B in reasonable comfort and time, it allows me to achieve more things during my day, it enables me to carry weights I certainly could not consider if walking or trudging from bus stop to train station, to the airport or taxi rank. And it certainly avoids the hassle of owning, feeding and clearing the waste products of a horse on a daily basis!

I did mention earlier about our differing attitudes to the ownership of a motor vehicle so I will blether a bit on that subject and try not to upset any who may recognise themselves in my descriptions. So where do we begin? Oh I know. First there is the young man roaring and revving his engine constantly day and night with little concern or understanding of the damage he/she is doing to their vehicle, or the amount of fuel they are wasting. Second is the *stiff necked* old man, driving a car much too big for his needs or capabilities. Why one would need an expensive top of the range Jaguar or Mercedes just to visit the shops or collect a pension? The stiff necked old men are called so due to their inability to turn their heads and make sure the road is clear before moving out into traffic, or switching lanes without indication. These old geezers know they are in the wrong but attempt to hide from the furious gestures directed towards them by other angry road users. By not moving their heads, they convince themselves they have nothing to fear.

Third in line are the illiterate drivers who appear unable to read road signs or directions. These drivers are the ones that cut you up on round-a-bouts as they have no idea which way they are heading. One time I did actually point this out to an elderly driver. He veered across the traffic lanes in front of me as we both came off a round-a-bout. He was in the wrong lane and so decided to cut across right in front of me in order to achieve his correct lane on the road. I hit my horn to make him aware that his move was inconsiderate and received many inappropriate finger gestures in return. Unfortunately for him, the traffic lights in front of us changed to red at that point and he was forced to stop. I do not normally suffer attacks of road rage but on this occasion I thought I would give it a go. Springing from my own vehicle I sped over to his and ripped open his door.

'What the hell was that?' I demanded.

'What?' he replied.

'You were in the wrong lane and cut me up!' I stated forcefully.

'No you were in the wrong lane!' he shouted back.

'Look at the road signs.' I said.

'No you look at the road signs.' He replied like all dimwits do, answer an accusation with an accusation with all the intelligence of a parrot!

I could see I would get nowhere with this plonker so decided to not waste any more of my time. Beside another driver who had witnessed the whole farcical manoeuvre enjoined the battle and also gave the plonker a few home truths. The lights were changing so I left them to it and headed back to my car, the witness did also but not after threatening physical damage to the cut up merchant who still simply repeated each accusation with the exact same accusation to the witness. Luckily I was working as a taxi driver at the time, so I spent the next three miles as close to the idiot's bumper as I could. I imagine the site of a huge black and battered old taxi hanging on his rear end may have caused him some concern, it certainly cheered me up!

Next is the most annoying of drivers, those who park across your drive or entrance stating they will only be a moment and there is no harm in them parking in that location for a short while. Of course it never is a short while!

And finally along comes the white van driver. He/she believe that the use of a white unmarked van entitles them to park where ever they like providing they have their hazard lights turned on. Blocking streets, entrances to homes and businesses, county lanes and they have even been known to obstruct entrances to hospitals! The very fact that they drive a white van appears to affect not only their ego but also their common sense. Maybe they should check the traffic laws of the country, hazard lights should only be used in cases of danger caused by a hazard or obstruction on the road that other drivers may not be immediately aware of. Hazard lights are not an entitlement to park anywhere, plus everyone can see a parked vehicle, simply parking is not a hazard even if it is in a damn stupid place!

OK, rant over. So the motor vehicle has come a long way over the last century or so. Across the civilised countries of the world, a motor vehicle of some description is owned by a huge percentage of the population. Our health and safety has greatly increased due to fire engines, ambulances and other rescue vehicles. The law of the land is assisted in its work by police cars and our coasts are protected with the use of rugged all terrain vehicles that can speed to the assistance of walkers and climbers who encounter difficulties.

Certain modern advantages owe their existence to the motor vehicle, like the school runs with small children screaming while encased in a massive four by four pimp mobile. Or the old men attending medical appointments in large and expensive gas guzzling vehicles. Supermarkets and Internet shopping sites delivering their wares straight to our doorsteps via a multitude of white vans. Relatives and friends call to see us before driving off to country pubs for lunch and shopping trips to the out of town DIY and garden centres. Not forgetting those lovely little disability scooters that resemble motorised wheel barrows and driven by those who have never driven before!

So kudos to the motor vehicle, there is not a problem with allowing technology to assist in making our lives easier and more comfortable. It is a pity some of these wonderful machines are allowed in the hands of some total idiots!

So now it is time to make use of my own motor vehicle, perhaps one of its main uses in life despite all the various possibilities and adventures such a vehicle can make possible. My use of this marvellous mechanical contraption today will be shopping!

Chapter Four. Shopping.

Observing people and trying to discover their place in life has continued to be an interest of mine. I am not admitting to being nosey, I just observe and wonder. From watching the kids at school I have progressed to observing people from the safety and isolation of my car. It is funny how when one is sat in a vehicle, the world around seems nothing more than a very wide screened television. Sometimes I even watch other people in their cars watching me! People also tend to forget that those outside the vehicle can see in as easily as the occupant can peer out. Thus it is not unusual to notice nose picking, scratching, fiddling, and tuneless singing or if there are two occupants in the vehicle, passionate petting! Most people though, when sitting in a car or other form of transport simply daydream, read a book or newspaper, listen to music or enjoy a conversation. I always have music of some form on in my car, silence itself can annoy me due to those strange sounds and noises my car makes that I don't wish to know about. I very rarely read when in my vehicle and having been married for many years, I certainly never take part in sexual petting, shame as it may be.

In these modern times, we all spend many hours cooped up in some form of transportation, how we deal with the time spent is down to each individual. I find it very relaxing to sit, listen to music and watch the world pass by my viewing screen, safe in the knowledge that I am in my own little world and no one can interfere. Of course this is not true, it is simply an illusion resulting from being in ones own property and hopefully protected by the laws of the land.

Unfortunately when actually driving, the illusion of viewing life through a screen takes precedence over the actual fact that one is moving amongst other steel clad contraptions, each with its own viewer encapsulated in warmth and comfort and each under the impression of seclusion and safety. This affects many people in different ways, consequently leading to a feeling of supremacy regarding other road users.

So, where is the best place to sit and observe people in their natural habitat and going about their own individual business? A supermarket car park is ideal, the bigger the better! Again speaking of life in the western world, most people have cause to visit a supermarket sometime in their lives, mainly through convenience, often for bargains and sometime just laziness or even entertainment if one has hordes of screaming, hyperactive children in tow. Whatever the reason, most of us shop in a supermarket of some kind. It may be a large DIY store or a general purpose and grocery store such as the big Asda or Tesco conglomerates. It may even be a furniture warehouse or garden centre, wherever there is a large gathering of people, observation of our fellow humans is easy and often unavoidable.

There is a chance that one or two of you may feel I am explaining actions that we all do in the course of our lives, and you would be right. I am simply discussing observations, meditations and conclusions resulting from these actions. Philosophers throughout the ages have ranged into deep discussion about behaviour and thought, some dwelling for months on the actual meaning of one word. Robert M. Pirsig, author of *Zen and the Art of Motorcycle Maintenance*

(1974) delved deeply into the meaning of Quality. Most of us have an understanding of what we expect quality to be; we seldom try to analyse how one would define or measure quality, we simply decide ourselves. If an item is better made, lasts longer or tastes delicious then we would state it had a good quality. For most people the thought process is straight forward, what we like and that what we do not like, people we like and those we do not. The truth is we all experience deep thought process sometimes, some more often that others of course, regularly depending on the situation or type of employment. An academic scholar may ponder issues deeper than a school teacher as he/she battles the gangs of teenage bandits (being polite here) who appear to have chosen schools as their 'patch' or battlegrounds. A surgeon may examine new scientific medical concerns in greater detail than your local General Practitioner, surgeons get paid more anyway, because they are a *cut* above the rest!

The same applies to all of us, at some time we all lapse into deep thought or speculation but do not realise that fact. I say most of us in the broadest possible sense, politicians and game show hosts excluded. If you were to stop and ask someone sat alone in their vehicle what they were doing, the answer with the highest percentage would be, *"Nothing!"* Many people do not realise their minds are busy attempting to analyse the world and its inhabitants as it moves past their car windscreen. Mention the word philosophy or sociology to most people and a glazed expression falls across their eyes or a muffled sigh or groan escapes their lips. They will claim they have no interest in philosophising and would never even consider it, but they love *socialising* . . . . !

So this is where they prove themselves wrong, in order to decide what one is interested in, one has to have given thought to the matter and without realising it they have philosophised, or equally possibly they have given a knee jerk answer with no thought at all! But allowing the benefit of the doubt to the majority of the populous, most would have given the matter some form of consideration before engaging their mouths.

Writing a simple letter or communication is another good point. Mentally debating the content of the letter and considering the actual words used and their meanings to ensure the full understanding of the reader. How one builds a paragraph and sentences and considers what subject concern or issues to include or discard. Letter writing is of such importance to many that they will ponder the contents and layout over many days, attempting to ensure the reader is kept interested, is not offended nor offence given to anyone discussed in the letter, unless of course it is intentional. Consideration is also given to the amount of information included in the letter, is there too much idle gossip, is the letter to be based on one subject, will it be in the form of a conversation, humorous, serious, sad, happy or indifferent. Writing a letter is a deeply philosophical practice that requires many facets of thought and mental investigation, examining the content, meaning and understanding to ensure the writer clearly makes his/her point to the reader. The art of communicating what is important in a given moment of a relationship, and putting it down on paper, has a basis in real soul searching. A chatty little note is one thing, but words offered in difficult moments, that help a friend when they need it, cheer someone when they need cheering, comfort when comfort is due, there is a real basis in philosophy in this simple task. We are all involved in the thought process of philosophy as we go about our individual lives, but few understand or acknowledge the extent of their mental activity when undertaking an otherwise lowly task such as constructing a communication.

Unless of course, one considers the mobile phone text language that is fast becoming the norm for most young people and a few older ones too. Text language consists of 'wot u doin m8t' and other such abbreviated rubbish that often defies basic logic. My mother-in-law is 78 years young and is an expert at shortening words and phrases when sending a text message. I am afraid I still attempt to text in full grammatically correct format, she can send several texts before I have completed one!

A very similar concept occurs when shopping, what to buy, what do I need, how much to pay and where do I eventually purchase my considered items. The philosopher Kant asked in his Lectures on Logic, *What can I know? What should I do? What may I hope?* Are these questions any more than what one would ask of themselves before undertaking a shopping expedition or writing a letter? Of course they are, but only in the depth of thought involved.

I do not consider my observations made while sitting in my car in a supermarket car park as deeply philosophical or sociological, nor do I consider them to be of any great value to others. In order to give a name to what I am seeing around and the depth I give to each thought process the term sociology as a label for my observations seems correct, although I hate to admit it. So perhaps a more suitable definition would be labelled as personal ramblings of a depraved mind! However I wish this script to be a simple record of my individual views on life, my thoughts and my observations. I do not aspire to lay claim to labels, classifications, titles or justification for what I am writing, but I would like to think some of my assumptions, thoughts and conclusions are recognised or identified by the reader in a familiar context. In other words, the reader would identify with items written and be able to state, *"Yes I have felt that,"* or *"I have wondered about that."* Or the final accolade, *"I know what you mean!"*

And so on with the show or shopping trip at least. Supermarkets in Britain are one stop shopping centres for groceries, DIY, furniture or flooring, I believe in the States they are called the same. Consumer items of every kind can be purchased easily, food stocks, toiletries, clothing, shoes, literacy materials, general household appliances and devices of entertainment such as televisions, DVDs, CDs, videos and including, books. Hopefully this one will be amongst them! The obvious attraction to shopping in such markets is convenience, no tramping from store to store, shop to shop along the high street, one journey suits all. There appears to be no class barrier for those who use supermarkets, rich and famous mingle with poor and destitute. OK some supermarkets are considered more up *super*market than others if one gets my meaning, and the more expensive the store is, the wealthier the customer. I may seek my groceries from Asda but I would not consider Harrods for a monthly shop, my meagre budget would simply not cope. Those clients may find food outlets such as Asda or Tesco beneath them but most would use either store at a push I believe, maybe not the famous such as film or music celebrities, they would be mobbed and harassed constantly by adoring admirers and grannies wielding trolleys!

In every day situations most people discover the need to frequent such an establishment. From a young man in a hurry who needs sustenance in the form of a quick sandwich. To the young couple with five very active and vocal children in attendance while doing their week's shopping. A more affluent couple purchasing luxuries for a dinner party and unconcerned about the final financial consequence, moving along the aisle alongside pensioners searching out the cheapest of the stores own brands in an effort to reduce the drain on their meagre pensions.

Shopping in such venues is often a family affair and can be seen, heard and on the odd occasion even smelt as hordes of mothers, fathers, young and older children, grandmothers and any other relatives that may be in the vicinity descend in hunting packs to search out the latest bargain. Groups of friends treat the weeks shopping excursion as a day out, especially the elderly. Others much like myself view the whole process with distain and attempt to break speed records in the rush to attack the shelves for our desired items and get the hell out of the place as swiftly and hassle free as possible. But for family, friends and other such groups of shoppers, a visit to a supermarket regularly follows a set procedure, first a quick investigation of the premise to seek out the possibility of even more relatives, friends or acquaintances that may be hiding in the bushes nearby. This preliminary search is then followed by a meal of fast food in the restaurant, and succeeded by a cup of tea or coffee and liquid sugar for the kids, just to ensure they have the just the right amount of energy to create as much havoc as necessary to fellow shoppers. When this ritual is completed with the obligatory pile of dishes, sweet wrappers and empty drink cans piled high on the now vacant table, the pack moves slowly off to make their purchases.

The actual shopping activity is set at a leisurely pace as each member of the pack closely scrutinises each item before purchase, while blocking aisles to other customers and seemingly unaware of their children happily demolishing a display stand. The group meander on their way while discussing new ranges and comparing individual tastes and expectancies with the others. Finally they achieve their objective and head for the final destination at the checkout aisles. Still in formation and still discussing matters of interest they form a formidable convoy at the till of their choice.

The main reason behind my many personal shopping misfortunes appears in the form of the elderly shopping gaggle. Pensioners, usually of the female variety but sometimes with a suppressed male dragging behind, who insist on walking in line side by side and thus taking up the entire aisle space and ensuring none can get past. Even the polite 'Excuse me please,' fails to achieve a result as selective deafness or turned off hearing aids allow the ancient ones to create queues in deliberate self denial of the growing line of frustrated shoppers forming behind them. To make matters worse, every ten paces or so, they all stop right in the centre of the aisle for a chat! Attempting to gently push past these senior shopping veterans can be fraught with danger as shopping trolleys, zimmer frames or steel tipped walking canes are brought into swift action and jab, crush or run over any stray foot that may have the audacity to forge ahead of the chatting coven.

These venues can also be a magnet for the single or lonely person, dawdling throughout the shelves idly examining products with no real intention of making a purchase. As they move round the displays they are seeking something else, company! Be it in the hope of spying someone they know for a chat or seeking romance and hoping to discover it amongst the Brussels sprouts. These people wander aimlessly about the store, desperately trying to find what they cannot simply purchase, some human company and a more fulfilled life. Young people and teenagers also seem to have developed a passion for large shopping venues, they too are seeking relationships and striving to find that certain person, or near enough anyway. Frequently it is just entertainment and a place to gather while examining the latest clothing and music fashions that are vitally important, still too young to visit a public house and sadly youth clubs and cafes centred on the young have all but disappeared over the last two decades. Cinemas, arcades and amusement parks have decreased in number and availability, funfairs have virtually stopped rolling from town to town and city streets have become hostile and dangerous grounds. So a warm dry and bright supermarket store has developed into the ideal honey pot venue to meet friends, discuss matters of importance and enhance any romantic possibilities in a safe environment.

Today I was forced under threat of things unmentionable to our local supermarket and as I stood seeking shelter on the end of an isle of shelves, I encountered a group of shoppers I consider to be quite rare. Six men paused near me, close enough to easily hear their conversation and the more I listened, intentionally I admit, the more amazed I became. These six men were obviously undertaking a task they were not familiar with, that of shopping! As they rested from the battle, a plan was being devised with all the competence of an army general. I concluded from their conversation that they were totally lost and confused. None had any real conception on the methods of obtaining ones supplies from a modern supermarket. So with earnest contemplation it was decided that they should divide the isles into sections and explore each one in detail before moving on to attack the next enemy shelves. The leader of this retail combat group signified an area of three isles that would be the object of their first sortie, and then he indicated the areas to invade next. Off they set in twos, following their brave leader as he ventured into the dark and threatening world of the grocery buyer.

I lost interest at this point though emotion welled inside me as I observed the shopping patrol move off. Watching this bunch of hapless men, I could now understand why men receive such ridicule regarding their shopping abilities. In truth I felt embarrassed, none of the group appeared to have any idea what to do, none showed the slightest common sense and ability to follow basic isle identification signs, all appeared slightly pompous and none it seemed, had the initial sense to consider making a shopping list! It was apparent that although they had been assigned to stores and supplies that day, they were sadly failing in their objective. I do not know why such a bunch of helpless men were attempting this task, possibly they had been sent out by their wives and partners to obtain supplies. However it was evident they were not local, possibly holiday makers roughing it away from the pampering city, or mayhap they were company employees on a team building exercise. What ever their reasons for being there, they were certainly out of their comfort zone. I passed them on a couple of occasions within the supermarket, still attempting to devise an outstanding plan with which to defeat the dastardly chore of shopping. Sadly they were an army without weapons as none had even the slimmest idea of what goods to purchase, their baskets were void, no bread, no milk, no eggs, nothing!

I hasten to add here that the vast majority of men are quite capable of undertaking the chore of grocery shopping with no trepidation but possible a little reluctance. This group of men were clearly the exception, their arrogance and ignorance sprinkled with large doses of male chauvinism only went to highlight their incompetence. I wonder if they are still there, wandering aimlessly about while continuing to devise intricate plans of action against this most aggressive of enemies, the shopping trip!

Over the centuries and the millions of retail venues I have been dragged through during my lifetime, I have made one observation that remains true whatever the establishment. As a hardened male shopper, I know exactly where to place myself out of the hustle and bustle of the supermarket environment. Like many men I position myself closely to the end of an isle section as near as possible to those strange unwanted items that reside at shelves end, a place of limbo while being advertised as on offer, half price or a new product. Here one will find the latest aroma in the gentleman's aftershave, always labelled with real manly descriptions that have no real bearing on the products capability at all. Otherwise it may be a new brand of potato crisps in garlic and old sock flavour, or boxes of chocolates still bearing the Christmas designs, an indication no one wanted them in June.

So I chose my stand with due consideration, attempting to pick the display stand offering the least favourite goods for sale. There I stand and await my next instruction or simply allow my shopping partner to bring the goods to me and place them in the shopping trolley. This method works well for both of us. I avoid battling fellow shoppers and my partner can keep track of my whereabouts. Observe this the next time you are reluctantly manhandled into a large grocery store, notice the sentinel gentleman like myself, standing alone while guarding a wire shopping trolley and positioned on the end of strategic isles. We are many!

However we do have a determined enemy, a foe who will stop at nothing to ensure we moved reluctantly away from our little area of relative safety, a non combatant who we cannot stop in their ruthless attack of the shelves. I speak of the elderly ladies, those female pensioners who insist on examining each and every single product on offer within the store. These formidable ladies are known to prey on the unsuspecting male, especially when no re-enforcements in the shape of a wife, girlfriend, partner or mistress are within the defensive perimeter. Spying an unsuspecting male standing quietly beside an end isle they quickly swoop. Pushing aside the defenceless victim, the elderly ladies push shopping trolleys, zimmer frames, walking sticks or ample bodies alone into the space between the male sentinel and the end isle shelves. Next they peer closely at each and every product upon the shelves, their expression shows they already knew the contents but the opportunity to strike a blow against the submissive male trolley pusher is too much of a prize. Once the scrutinising is complete, these formidable women vacate the end shelf vicinity, forcing the male further away from his favoured venue of seclusion, before scurrying away in their hunt for the next victim.

I classed myself as a hardened shopper because these days I find I cannot allow this form of retail bullying go without some form of retaliation. So these days my shopping trips have become moderately more interesting. I seek out my stand and await the first onslaught from the female pensioner. I allow myself to be ushered away from the end shelf without complaint. However as my antagonist completes her search of the end isle stock and begins to head off in my direction again, obviously attempting to prolong the victory, I snap into action like a Ninja warrior on a suicide mission!

'Don't you want one of those aftershaves love? Looks like you could do with a shave!'

Or, 'Crisps not to your taste after all? You can suck them to death you know, don't need teeth for that!'

These and other comments spill from my lips before I rapidly vacate the immediate area, and place space between myself and any possible physical violation by walking canes, handbags or thrown husbands. A small victory to me!

Sitting in my vehicle and watching my wide screen viewer outside one of my local supermarkets, I am constantly drawn to form conclusions and opinions on those people I see venturing in and out of the store. Who are they? What do they do? What kind of person are they? These and many more questions invade my mind in the safe and secure warmth of my car. When I discuss these questions with others, I frequently get blank looks, a scratch of the head or I am fended off with a blithe answer. All too often these days it seems, people just do not take notice of who may be around them, many will know what character is in what TV soap opera but have no idea who their local councillor is or even their immediate neighbour at home. They may have a rudimentary knowledge of world events but often this knowledge is unclear and leads them to form inappropriate conclusions. It is the same with observing life, request someone to describe another from memory and the answer will frequently be vague. Identified as tall with black hair or short and fat is generally the limit of a description, a vague verbal image that embraces a very high percentage of the human race. On police or crime programmes and films, the officer always manages to extract an excellent likeness of the villain, in truth however I suspect this rarely happens and the law enforcement agencies are left frustrated and impotent.

    Observation in general appears to be a dying art, often it is through fear. We avoid eye contact or not invading some ones space in order to defuse any possibility of confrontation or danger. So the general populace of our civilised world habitually fails to see or note the events or people in their immediate environment. Please do not presume I am placing myself in a superior position, I enjoy observing people and things around me, forming opinions and conclusions that may in deed be very far from the mark. This activity and the observation and conclusions I reach are the basis for this book, simply put, I am writing about what I see.

For example it appears that different types of people shop at different times during the day. I went out early this morning, gathering Christmas presents in a vain attempt to beat the hordes and I observed a couple of details of interest to me. The first observation was that large numbers of men appeared to be shopping on their own. With their heads down they race into the shop and exit again with all due haste a few moments later. No hassle, no waiting, no pondering over each item considered for purchase. Just in, buy and out! Why would single men, I say this because I recognised many of them and know them to be single, prefer to shop early on a Saturday morning? The answer could lie in the individuals wish to avoid queues, beat the crowds of late risers who appear to have all the time in the world as they stroll round each store with numerous children, friends and relatives in tow. Or perhaps it is the result of good planning and knowledge of where and what to buy and why. However there is usually the man who appears to be preparing for work and is gathering his working lunch, newspaper or alcohol, depending on his chosen form of Saturday employment.

Certainly on a cold wet Saturday morning, only the busy or hardy customers appear to venture out in search of a bargain or necessities. The reason may go deeper, shopping has always been a hateful experience for most men, and some may even consider it is not the accepted task of a *real* man! Man the hunter degraded to man the shopper. Is there a challenge in simply entering a shop, making a purchase and leaving? No; not really, so perhaps no challenge equals no enjoyment or interest, just necessity. Deeper still could be the solution, independence, a desire to demonstrate a capacity to keep ones life in order and to survive on ones own. Perhaps uneasy memories play a large part in the reasoning, making shopping a suitable challenge after all? Of course in many cases the reasons and objective will be more superficial, get the shopping out of the way before the pubs open or the football starts! However I would theorise that in many cases all of the above mentioned are inclusive and that the necessity of shopping, of gathering food to survive, of facing and interacting with strangers, of travel, of finance, and of living, all play a part in the minds of those men descending on local venues of commerce in the early hours of a weekend morning.

I noticed the women as well, obviously! But what I mean is that many of the women also appeared to be on their own, smartly dressed and with the appearance of rushing off to their individual places of employment. Others scampering round the shelving lines with determined expressions fixed upon their faces. They know exactly what they want, where it could be found and what it would cost. Anyone obstructing the speedy progress of the efficient women was met with a stare that could freeze a soul into submission. While most men hate any form of shopping with gusto, women tend to consider it just another chore that needs to be attended to, no fuss, no objections, just get it done. I too prefer attacking the retail establishment during the initial hours of the morning, and I can fully emphasize with both male and female reasons for the early shop, because by ten o'clock the hordes, family groups and ancient anarchists begin arriving!

The next observation I made was the faces of those people, both men and women as they strode quickly amongst the aisles, faces full of concentration, of contemplation and purpose. Each face failed to extrude the humanity, warmth, feeling or true mood of that person. There appeared to be little cheer in their expression, poker faced is a term I believe may fit the description. Is it an underlying fear of confusion, of danger or fear of the day ahead that haunts these people? Is it that they are merely preoccupied with the task at hand or perhaps only the early hours that prohibit feelings of generosity and compassion? I realise many are not what is often termed *morning people*, preferring to remain silent as their minds increase in activity after the initial waking, still tired and weary after a fitful restless night and an insufficient amount of sleep, perhaps even the result of a very good night on the town or at a party. I frequently observed people I recognise behaving in the same detached way, even though I know them to be kind, generous and garrulous at other times during a day. A brief *"Hello"* as they speed past, no interest in conversation, polite or as friends, no real eye contact or acknowledgment, just the one word as they continue purposefully on their way.

Speaking from experience and although I am a morning person, annoyingly so to many, I am in fact a whole day person, cheerful and active(almost) from the moment I get up until I return to bed at night. However I do find I ponder any worries or problems deeper and for longer in the evenings. All the fears and worries from the day flood my mind and it has become something of a strenuous mental activity to subdue them in order to face new challenges in my life. On many occasions a problem pondered as I retire to bed at night is still racing through my mind as I awake. This situation is worsened by the fact that I have been unable to sleep because of the worry. In times such as these, I would appear as those early shoppers, confused, reluctant, and afraid, devoid of the basic ingredients of humanity. But these are rare occasions for me and could not be classed as a constant factor when attempting to analyse the reasons behind the faces of the early shoppers.

It must be accepted that people do not shop early simply as a result of worries, problems or a poor nights rest, other explanations should be sought and considered. Regret, disillusion, disappointment or sadness concerning an event, argument, action or response from the day before will play heavily on many minds in the clear light of day. Something said, something done that caused hurt to a loved one, a wish to take back a comment, remark or insult that resulted through a brief bout of temper, sullen mood or mistake. An action that should have been avoided or action that should have been taken, an event that did not happen as expected. Any of these regrets, worries and concerns can be seen clearly in the morning of a new day as remorseful and possibly hurtful to a person or persons not at fault and innocent of your mistake. A petty argument with a friend or partner that results in a break up the night before seems unjustified and unwarranted the following day.

Of course it could simply justify the belief that we are, as a human race, real grumpy and unsociable in the mornings. Nothing more serious or deep rooted than the desire to have an intake of caffeine or we are just extremely annoyed at having to vacate our warm and comfortable beds on a Saturday morning! There are many,

many conclusions that can be reached in an attempt to explain these faces wandering the shops in the early hours of a Saturday morning. No one will really ever know for sure, I believe each and every one of these conclusions will relate to someone, some where and of course there are many more I have not considered. The question remains, what lies behind those faces of people shopping in the early hours of a weekend morning? But more importantly, who cares at that hour of the morning?

Of course these days, the larger shopping centres offer 24 hour opening, thus enabling those who work shifts, nights or to assist those who commute long distances to obtain supplies at a time more convenient to their life style. This opportunity has not gone amiss by other shoppers, those that desire peace and sanctuary from the average hordes of shoppers. Many now prefer to shop `out of hours', leaving them free to move round the store, examine each product at leisure without being bumped, pushed or having a small child gnawing on their ankle. Possibly this change in shopping habits could benefit those early morning faces. Of course it is likely they are aware of 24 hour shopping but choose not to do so or cannot do so. But night time shopping can be a whole different experience. People appear more relaxed, less concerned with time and are more likely to speak or acknowledge each other in passing. Once the usual small groups of young men and women visiting the store searching for further alcohol after closing time in the pubs and bars has subsided, the store itself becomes a quiet almost reverent place. Night shift workers go about their chores stocking shelves and clearing the debris from the day with as little noise as possible. A practice likely deriving from a life times conditioning of being muted during night hours, *sleeping* hours. It can be noted that people automatically lower their voices during a conversation or discussion that takes place in the very early hours of the morning, as if they feel a raised voice would awake those still sleeping in the communities nearby. Loud high heeled shoes and jingling jewellery are shunned, as are clattering boxes, clinking bottles and rattling tins. Staff and customers go about their routines as quietly as possible while keeping an appearance of norm. This cathedral like atmosphere contributes to the hushed actions of those working or shopping late into the night.

Once when visiting a local store very late at night, I chanced upon a friend in an aisle and during our short conversation I asked him what his thoughts were on 24 hour shopping and his answer surprised me. He stated the necessity of shopping frightened him, all he saw around him was danger and confusion, he constantly felt 'something was going to happen any moment' and could not remove from his mind the image and fear of the whole shopping experience.

'I see the whole picture,' he said, 'and in 20 to 30 years things are only going to get worse!'

Unfortunately I had neither the time nor the inclination to seek further explanations to his concerns but of course I can surmise a conclusion from his choice of words. When undertaking the chore of shopping, especially in a large city centre, many people experience fear, uncertainty and apprehension when faced with mixing and interacting with such great numbers of complete strangers, unfamiliar surroundings and confusing venues. Modern cities, shopping centres and malls are filled with people of all ages, creeds, religions, race and prosperity. Those with an underlying fear of humanity see muggers, thieves, cheats, frauds and bankers behind every corner. Young people hiding their faces under hoods and baseball caps and giving an impression of violence when all they really intend is to follow their own form of fashion. In truth shopping is not a frightening occupation for the majority of consumers, most see no bigger problem than deciding how much to spend and on what. Danger is there of course, many elderly people, women and frail people do fear to be out during the evening or night hours, daylight at least offering some security against the growing violence in our main cities and towns. Even in many rural areas people dread being out after dark for fear of roaming gangs of youths, drunks, addicts and Time Share salesmen. But hang on a moment; has there not always been an unscrupulous factor in every society? Yes there has, so why so much fear today? It cannot be that we have all suddenly become paranoid, en mass afraid of the monster lurking on every corner, surely not. Perhaps we are simply more aware of such things thanks to the power of the media, our prized televisions telling us about the troubles of the world every day as we sit in the comfort and security of our homes. Modern police programmes and the onslaught of celebrity hosted crime programmes are not really designed to frighten, but many do as our brains turn to mush in the presence of an orgy of egos, self obsession and arrogance. I will admit that programmes such as Big Brother frightened the hell out of me, along with the X-Factor, The Voice and I am a Celebrity, get me out of here! Any one of these so called entertainment programmes can scar a soul for life!

In the past stories of violence, robbery and murder relied mainly on word of mouth and consequently fewer people became aware of these facts. Villagers often heard little of what went on in distant villages or towns, and people in rural areas seldom took notice of happenings in the major towns or cities. So perhaps there is no greater threat, no greater reason for fear when undertaking the shopping excursion? Perhaps it is just our natural instincts for survival that make us weary of strangers and places unfamiliar, maybe it is all in the individuals mind? I cannot answer these questions, is today's society more violent or are individuals simply more conditioned to acknowledging the possibility of danger in their environment? Personally I feel it may be a combination of both. In the melting pot of the human race most people have the right to choose how they live their lives, their ideas and conceptions, their fears and concerns. People shop at night through their own choice, what ever that may be, most shop during the day, again their choice, often the time we shop depends of our lifestyle, work hours or commitments. Most of us do not plan our shopping trips round concerns of safety or to avoid possible threat, but it must be noted that some do, for their own reasons and assumptions, their own fears and apprehensions.

A simple trip to the shops brings out many characteristics in all of us. Those of us who see going to the shops as a bore, or a chore that must be done, a necessity, or even a minor inconvenience may wonder at the fears of the man in the 24 hour supermarket, but who are we to judge? I hate visiting the dentist and suffer attacks of anxiety and panic just at the thought! Taking a car for an M.O.T can be extremely stressful to those on low income, fearing a large repair bill, even visiting relatives can be a cause of concern to many, especially those weird in-laws! Why should shopping be any different? We all have to undertake tasks we dislike or fear, leaving our secure and comfortable homes filled with people and objects familiar to us, and to many it seems, shopping does come into the fear category.

In general though, shopping is neither threat nor a danger, it is just a modern method of survival. Man no longer needs to test his wits against wild animals, spend hours searching for roots and berries, fighting rival tribes or enemies in his quest for food. Now food is found on shelves in large or small retail establishments, the customers only concern now is what they can afford to buy. So whether we love, hate or fear to shop, obtaining our basic needs plus the odd luxury relies on a trip to the shops, until the growth of internet shopping of course, but maybe that's material for another chapter? Internet shopping! My idea of heaven!

As I sat in my car watching the world go by as usual one day this week, I became fascinated with the individuals response to shopping and this lead to the chapter you have just read, hopefully. Multi cultural, multi religion, multi race, rich and poor, fit and sick, young and old, all eventually pass by the wide screen viewer attached to the front of your vehicle. People with mounds of bags and purchased goods alongside the single item purchaser rush out of shops and stores around the world, every minute of every day, no wonder some of us get hang ups about the task of shopping. Everything one needs to survive can be found on a shelf it seems, food, medicine, furniture and of course, fashion!

Chapter Five. Fashion

When discussing fashion one has to take care, observing people as they go about their individual lives can initiate strange chains of thought and personal conclusions are reached, but are these conclusions or assumptions correct. Questions arise about how they might live their lives differently to others, or for that matter similar to others. Many say they are not followers of fashion but is that strictly true? All know what clothes they like to wear and know what image they want to portray in public, therefore does this indicate we all are all followers of fashion?

Each and everyone of us will strongly state that we do not care what is in fashion and what is not, especially we men, we all like to proclaim our individuality in our choice of clothes, but in reality do we not simply follow the herd? Do we really not care how we appear in the eyes of others? I think not, how we appear to those around us is highly significant. We may not be bothered to make an effort, but I believe we all care in some small way about how the world sees us.

Today's teenage fashion, and I am including this because it depends on the moment in time when you are actually examining this text, seems to me not unlike that of the 1960's. Girls have started wearing hipster jeans and trousers, tight short tops that expose their often ample midriffs, and hooded jackets similar to the 60's Parker coat. Girls' trousers have also become very tight in the leg, as in the 60's, and are often worn with a wide belt. Trouser bottoms flare out once again, though not nearly was wide as those worn in the late 60's and early 70's, thankfully!

The mini skirt has made a huge come back amongst young girls, in some cases unfortunately. Yes I realise what I have just said but I still feel somewhat uncomfortable when a girl of not more than 13 years of age insists on wearing skirts extremely high and revealing. What happened to their youth and their childhood if they are already making themselves noticeable to boys? Obviously a mini skirt can be very attractive, but on a young woman rather than a young girl, or maybe I am just getting too old! I have always been of the opinion that what a man cannot see is often much more appealing than what he can see, his imagination will fill in the parts left obscured, frequently in more flattering detail. Seeing too much flesh can be a right turn off, wobbly belly buttons flopping over trouser tops, *love handles* looking more like tractor tyres forced under skimpy tight T-shirts and the *builders bum* now has a rival in that of girls *charming cheeks*! Both usually unaware they are making an arse out of themselves!

Anyone in my lowly opinion can heighten their appearance and therefore attraction by just wearing clothes that actually fit, and if that means changing the style of ones fashion then so be it. Following the herd in fashion can cause one to resemble a buffalo rather than a peacock.

Men fall into this category just as much as the women. Out comes the sunshine and off come the shirts, normally revealing a beer barrel waist rather than a 'six-pack', badly chosen tattoos or skin so white one needs sunglasses! However if the man decides to remain somewhat clothed and can be considered a large man, then out comes the bright yellow sleeveless vest and multi-coloured shin length shorts, small white socks and dirty trainers or sports shoes.

Young men today, and I mean mostly teenagers have developed a fashion of their own. I speak now of those young men who insist on wearing their jeans below their bum, showing the world their often none too clean underwear. This is especially distasteful when the young man walking the street in front of you shows obvious signs of having devoured a particularly strong curry the night before, which has resulted in the common side effect of an upset stomach - and bowels!

Men are frequently and insistently being told they do not have any fashion sense and that they need a woman to dress them, figurative not literally! Many men play on this misconception, mainly to earn brownie points with their partner but in truth this is not the case. Men dress how men dress and always have done. If 100 women were to give an opinion on one man's attire, he would receive 100 different pieces of advice concerning his choice of clothing. It is not that men are not fashion conscious or unable to match simple items such as a T-shirt and jeans, or trousers and a shirt. It is more the case that they have chosen items of clothing outside the fashion criterion of that particular female judge. This is not fashion, it is personal choice. Changing the angle slightly, how many women seek advice on clothing from their male partners? Not enough! I am not having a go at women here, I want to live a little longer, I am simply stating that just being a woman does not make one a fashion expert. Merely look around and take note of the women forced into clothes way too small for them, those with *muffin tops* or the larger *cake shelves* otherwise known as protruding stomachs hanging out at their waist line. Consider those hideous leggings worn by many of those carrying a tad more weight than they perhaps should. Or those squeezed into skirts that are far too small for the ample body features. Of course I suggest caution here, if caught peering at a woman's clothing or even worse, criticising her chosen dress, it could be hazardous to one's health! So gentleman, wear what you like, woman wear what you prefer, do not dress just to please another's expectation or personal opinion of fashion. However please remember that not every woman is a fashion expert and not every man is a slob!

Shoes are an item of fashion that has changed. Gone are the dainty light weight stiletto ladies shoe of the past, instead modern footwear tends to be thick heeled, thick soled chunky things that often look much too heavy for the wearer, especially when the wearer is a young slim girl with a very short skirt. I often have the impression of thin sticks with large blocks of wood attached at the ends. Cynical I know, but as I have said before, these are my own impressions as a biased observer. A friend of mine once expressed his preference for girls in tight jeans and big *Doc Martin* style boots. This was his personal idea of an attractive fashion at that time. He married not long after issuing this statement and it has been noticed that his wife has never been seen in tight jeans and large boots!

I mentioned earlier that each individual likes to state their originality in their choice of fashion, well to back up this statement I actually asked several girls/women for their comments and I have listed them below.

1. I wear what's comfortable, I love soft stretchy fabrics. I cannot stand wool it is scratchy but makes me look good. In that order. Comfort trumps style. Unless it is a special occasion & then comfort goes out the window (pantyhose & high heels were invented by Satan!)

2. I wear what my friends wear, it looks good and is cheap to buy, I do not go for designer stuff.

3. Comfort first, beauty second. But perfect clothes, to me, fill those two requirements in one. It is also a statement, clothes tell things, so what mine tell I think is that I care about comfort, and I care about beauty.

4. I prefer to be casual usually jeans & a sweater or casual top. I have to dress up somewhat for work. I do not mind wearing skirts once in a while. I like a long skirt with boots for winter. I love turtlenecks because I am always cold. I like to look stylish. I am budget conscious so everything I wear was a bargain. I do not have to wear designer labels. I really couldn't care less about that stuff. I tend to wear things that are form-fitting because I am proud of my figure. As a kid I was a stick figure and I wore everything extra large. I think I was hiding in my clothes. Becoming androgynous. As I developed curves I started to dress more feminine.

5. I used to dress strictly in black. I didn't own anything that wasn't jet black. I was a Goth. I loved the music, the bars, the clothes; the whole culture was appealing to me. I have become a little

less dramatic as I have gotten older I suppose. No more black nails & skull earrings. I used to wear rings on every finger. Now I feel strangled by rings. Well, I suppose I could tolerate wearing one ring (platinum diamond perhaps), if it came to that ha ha...!

6.   I wear what I like. My clothes are a reflection of me. Casual, sensual, down to earth, unpretentious, occasionally dramatic.

Only one of the above statements admits to wearing the same style of clothes as her friends, most have perfectly good reasons for their choice of clothes and are happy in that choice. However I still maintain that most if not all of us follow a fashion of some style, possibly one that is expected of those in a certain age group, particular profession or even a salary bracket. Few of we slightly more mature people would want to be seen in a very short skirt, tight jeans or bedecked in chunky garish jewellery, and showing our belly's to the world. Well certainly not me, I would look even more ridiculous than I do already! Of course it is not only girls fashion that has changed, some men's modern choice of clothing often leaves me amazed!

When it comes to fashion, the modern young male contrives to fit into one of four groups; the tanned and bleached hair of the surfer look, the hoodie and hanging jeans of the street look, the simple t-shirt and jeans giving the casual look of the wannabe ordinary guy and the shirt, jeans and gelled hair favoured by the lad about town. The first being that of the young man who appears to enjoy life to its full, wearing simple T-shirts and shorts or even the new three quarter length jeans, simple trainers and a healthy tan. In my observations I have noted that young men dressed thus are more inclined to avoid trouble, often have good jobs and stunning girlfriends. I cannot of course say this is a true statement; it is simply something I have observed. Putting this theory to the test, imagine walking down a dark alley way late at night. Are you most likely to meet a young, tanned guy with a guitar, football or surf board under his arm?

The second is the hoodie wearer, showing the world a sullen individual trying to hide his identity, or his feelings; his own fears or love or hate under a 'hoodie' as he slouches in a corner of the alley? In third place we see the ordinary or perhaps nonchalant man who wears t-shirts and jeans and has no real interest in clothing so dresses for convenience. This style of dress is by far the most commonly seen in all ages of men, though the choice of t-shirt slogan can be a subject of personal opinion. Finally in staggers the gel haired, deodorant smothered individual out on the town in search of conquest. By day the individual may adopt the casual t-shirt and jeans attire; but come the evening, a metamorphic change occurs and t-shirts are swapped for dubiously patterned tight short sleeved shirts and designer jeans or trousers as the casual man becomes the lad about town.

Yes I know I am stereotyping here but it is only to get a point across. Fashion trends not only reveal a style, they can also give an indication of the path that person has chosen to follow in life. Good or bad, concerned or indifferent, happy or sad, rich or poor, fearless or afraid. Many of these emotions or persona can be reflected in what an individual chooses to wear. Alternatively the choice of clothes may be deliberate, the wearer is trying to communicate to others what he/she feels about themselves.

OK, I can imagine many of you screaming your opinions into this book, and many of you will have a far better understanding of this dilemma than I, but as I cannot hear you I do not know what you are saying thus I still have no real answers. Or perhaps I do, perhaps the self styled free spirited individual is not such an individual after all? Analysing these thoughts as they appear on the page has caused me to examine my own sense of individuality.

So are fashion choices deliberate after all, does the young tanned guy appear as he does simply because that is the impression he wants you to see? What about the young man dressed in a hoodie, what is he trying to tell us? Again alternatively, is he trying to hide something? Is he concealing his true fears, concerns, lack of confidence or shame? Or is it simply that he likes the printed logo adorning the front of the garment, or because the garment is warm on a cold night, possibly it is purely that the hood is keeping the rain off his face? I would consider the latter to be much more probable. We cannot know why others wear what they do, it may be deliberate, and it may be from necessity, convenience, comfort or a favourite item of clothing.

Judging people by what they choose to wear cannot be an accurate form of assessment. People wear what garments they decide fit at that apposite moment in time. Girls will wear short skirts when the situation deems it, young men will wear hoodies when the moment demands it. Business men do not spend their entire lives in a suit; doctors do not wear white coats when relaxing at home with their families, in fact a doctor in a white coat in the UK appears to be a fashion of the past. Many doctors dress so casual that hospital security guards never know which person to throw out on a Friday night. The visual difference between a young male doctor and one of the lads suffering after an inebriated night out with his mates is negligible.

But our style of dress is also often a work uniform, a mechanic would not frequent a restaurant in overalls, and circus clowns would not go to the theatre in their false wigs, huge red noses and oversized shoes. So why do many of us still attempt to judge a person by what they are wearing when we see them for the first time? Clothes are another expression of communication, nothing more, telling the world what we want them to think, feel, assume or judge about us in that fraction of time. Be it in a suit, jeans, short skirt or overalls, clothes are just a visual form of statement, nothing more.

Earlier it was mentioned a person wore black because at that time they considered themselves to be part of the Goth fashion. This is a clear statement; they wore black to signify their interest in that style of dress, music and life style at that time. Others have stated that they wear clothing based on comfort reasons. Yes most consider the comfortable garment in the first instance, but what does one chose when the situation demands other than casual? Certainly jeans and a T-shirt can be categorised as comfortable and casual, but would you attend a job interview dressed thus? No, you would dress appropriately for the occasion, assuming you actually wanted that particular job.

Some suggested they dress in relation to their finances. Well of course we all do that and wearing designer clothes is not easy if you have not the funds to purchase them, we all dress in what we can afford. The statement that, *"I wear what I like."* is also untrue, we wear what we have to for the situation we are going to be in, we may try to convince ourselves that what we are wearing is our own individual choice, but is it? No of course not! If I wore what I wanted for much of the time, I would never be out of my pyjamas and slippers! Sloth being one of the seven main sins and I am guilty. One woman wrote; *"My clothes are a reflection of me. Casual and sensual, down to earth, unpretentious and occasionally dramatic."* Well sorry dear but I would consider this very pretentious! This is not an indication of who you are, it simply demonstrates how you want others to see you. If you are unpretentious, why tell others? Your clothes are not a true reflection of yourself; they are a reflection you wish others to see. I may dress like batman but it sure as hell does not make me the Capped Crusader! Just some idiot with a cloak and his underpants on outside his jeans!

The retail fashion industry changes purposely to ensure job security and profits, obviously, it is a business after all. The industry has been known to change over to a completely new and opposite style and design over night almost, so that other designers and fashion artists will continue to spend money in order to keep up with fashions. For example if flat shoes are in this season, they will purposely designate the extreme opposite, possibly 6 inch stiletto heels to be in style the following season to keep the market buoyant and profitable. Fashion merchandisers and buyers set the decisions for what is in style. Designers typically follow what has been predetermined at fashion seminars, hosted by merchandisers and buyers, to be in fashion the next season.

So are fashions important? Yes. Do we deliberately decide what to wear to fit the occasion we are to be in? Yes. Are our clothes a true indication of who we really are? No, I do not think so. We wear what we think others expect us to wear and when we think others will wear similar. Yeah I know, that sentence made your head swim but I hope you will get the gist. We do not freely decide what to wear, it is governed by many external factors. Who we are going to be with; what we are going to be doing, where we are going, what we can afford and what impressions we hope to give to the world around us.

OK enough of fashions, I just hope I have not offended anyone in the course of my ramblings, but this is how I see the world and those in it. I am not a fashion or clothing expert and I may be well off target in many of the things I have said. However I did surprise myself, and I am hoping that if you are honest with yourselves, you may have a surprise of your own.

One final thought, there is one vastly over-ridding factor that dictates what we wear and when, a factor with a more powerful allure than the gang we wish to join, the club we belong too, the profession we chose or the future we see for ourselves. Well I expect you have come up with ideas of your own by now, and the majority of you would be correct. We dress to attract! We dress to impress those who appeal to us the most; we dress ourselves as a bird displays its plumage to find a mate. Yep I know there will be certain members of our society shouting that I am wrong, but I am not. Be you heterosexual, bisexual or homosexual, each and every one of us desires the company and contact of another. So please do not shout indignant denials or accusation, attraction is the underlying force that drives fashion, along with profit of course.

Chapter Six. A Quickie on Attraction.

As I have just concluded a chapter on fashion, it would seem appropriate to proceed with what is considered the true driving force behind our choice of fashion, and that driving force, that overwhelming desire to attract and impress, that icon of evolution, sex! But do not panic as I am not going into intimate details of sex, I intend only to discuss some of my observations on human attraction, and of course much of the comments and observation recorded here will be based on personal experiences. Also it may not be polite or good for ones health asking strangers about their sex life! I admit this chapter is somewhat risky for me, this is not an area that one can discuss freely without offending some of the more delicate and sensitive amongst us. I assume most readers will have some form of understanding and experience about the desire to reproduce so I will not linger on the act of sex itself, instead I will attempt to narrate some of the experiences and observations concerning human attraction.

What is it about another person that attracts us to them, is it a perfect body? That body one sees flaunting itself about on television shows and adverts for slimming aids. The oiled supple bodies showing not an ounce of fat or those unhealthy sized zero women who constantly remind me of a childhood saying. As my friends and I reached gently into the first stages of puberty, we developed an expression to describe those very thin girls whom nature had not yet moulded from a girl into a woman. This phrase was '*Lucky legs! Lucky they didn't break off and shoot up her backside!*' Yes I realise this is a cruel expression but we were very young males, attempting perhaps to display to the world our deep understanding of sex and attraction. Unfortunately it actually highlighted our inexperience. The phase existed though and it constantly returns to mind when I ever witness a young woman striving to obtain that walking skeleton figure. It is horrible! Good grief girl, get some curves!

So what defines a perfect body? Each and everyone of us has our own conception of our ideal perfect body and at this point I have no doubt many men are reacting to images of large breasts or pert bottoms flashing through their minds at this very moment. While women may be dwelling on large biceps and flat stomachs on a young bronzed man. In truth while these factors are obviously appealing, are these really what we all seek in our ideal perfect partners and ourselves? If so I have failed miserably! Do we consider intelligence or a sense of humour? Certainly most of the `Lonely Hearts' adverts state a good sense of humour is important but can the ability to laugh really be that important in maintaining a healthy and loving relationship? Can one really picture circus clowns as the sexual studs on our era? Pies in the face and water balloons everywhere! What about wealth? Does this play a large part in what we may consider attractive? Possibly a strong provider could be deemed a greater factor, is the ability to raise children successfully a trend searched for by men? I seem to remember research done long ago which stated that a pear shaped woman was considered the best shape for child producing. Obviously I cannot make any further comments on the statement simply because I have not a clue! Pear shape, apple shape, short, tall or outlandish shapes to me mean little but possibly do have some influence on what some may consider attractive.

Other items may also have a bearing on who we decide is attractive, fashion, do they dress well? Posture; do they hold themselves correctly? Conversation, do they embarrass themselves and others about them as soon as they open their mouths? All these factors have a significance in what we regard as attractive, but is there an overriding factor? And if so, please tell me what it is! Is our attraction based on something more fundamental? Do we look for a pretty face, nice hair or teeth or are we subconsciously seeking deeper traits that at first elude our eyes?

Personally I have been fooled on many occasions, choosing a vision that immediately lights up my life, creates a physical reaction that can be embarrassing in public and sets daydreams on a rampage of fantasy and imagination. Phew! How wrong can one be? But will I learn from these experiences? Probably not as like all others, my natural instincts or perhaps a better description may be bodily functions take over and lust will rear its head. I remember as an over sexed young man in my mid teens falling head over heels for a girl with long black glossy hair, huge dark brown eyes, full red lips and a figure to match. I pursued this vision of the female variety for months, plying her with compliments, small gifts and constantly attempting to ensure she noticed me at every occasion. I really fell heavily for this girl, she became the most important consideration in my inexperienced life, and I could not see my future without her by my side. I have no misconceptions that I made a total arse out of myself, all my friends began calling me her puppy dog but I did not care. I needed to make this girl my own and was willing to withstand all forms of ridicule to achieve my goal.

Well one day I did! I finally wore down her resistance and we became an item. I could not believe my good fortune as I sat holding hands with the vision of my dreams. I desperately wanted to kiss her but held back in the fear of destroying the moment. In the youth club disco hall on which we all converged two or three evenings a week, I sat on a hard wooden chair on the edge of the dance floor while the disc jockey blasted out tunes and an incoherent dialogue as his show progressed. I wormed my way as close as physically possible to the object of my attraction, ignoring the flashing lights and the inane rubbish emitting from the disco. I even remember what she wore that night. A woollen yellow turtle necked sweater with a kilt skirt just above her perfect knees. Large gold coloured and inexpensive hooped earrings hung down, peering from behind her ebony locks of long flowing hair and bangles upon her wrists. A gypsy queen with classic features and a ripening figure that suggested her attractive looks could only improve with time. Quiet and aloof, holding my hand in return and not rejecting my moves to edge closer, but not encouraging them either. For a few moments time stopped and my heart swelled with desire and pleasure as I sat on the hard chair with a goddess beside me. But that moment lasted but a heart beat and things between us began to change instantly.

This girl, whom I pursued, longed for and of course lusted after suddenly lost her charm. Not the physical charms you understand, they all remained prominent in my vision, no it was something deeper that I had not noticed before. In my continuing attempts to woo this beauty I had failed to notice the lack of rival suitors for her attentions. Why had such a beauty not attracted more attention? Why was I not fighting off other randy young men with a better physic, more money and especially those who actually owned their own motorcars in stead of just two limbs with worn shoes stuck on the ends such as myself? Not one boy ventured near this girl and I can categorically state that I was not the reason they stayed away. Young men in the full rage of puberty and manhood rarely consider the consequences of stealing another man's female companion, they are lead by something primitive, and usually a small organ leads the way! Need I further explain this detail? No I thought not!

As I sat in the lap of pure sensational pleasure I began to notice others details that previously appeared absent, not just the lack of rabid young men queuing up to take my place at her side, but other, some more subtle concerns. I eventually realised that although we had been sat snugly together for some time, no words had been spoken, least not by the girl. Nor had she made any move towards me or showed an indication to be closer. Nothing! She simply sat and deemed to allow me to grasp her small white hand in mine. To this day I wonder how she may have reacted if I had suggested we went outside for maybe a tumble and a frolic, or even simply placed my arm around her shoulders. It is with some regret that I attempted none of these things, after all I was a full red blooded young man myself with all the usual stirrings that arose in moments such as this. But I was still quite young so did nothing, not because I did not want to but that I was convinced I would not prevail in any way, shape or fumble. Deep inside I knew if I even considered taking the relationship a small step further, the only step I would receive would be her foot on my head!

There we sat, a frustrated able bodied male with a healthy fixation on a beautiful girl, who unfortunately had the soul of a fish and the vanity of a reality programme celebrity! That was when I realised my mistake. This girl had no feelings towards me or any other boy, nor was she attracted to girls in a same sex situation. Nope, this girl had only one desire in her life, one attraction and one huge flaw. The only person this girl was in love with was herself! As this realisation hit me I noticed she had chosen seats near a wall mounted mirror, placed there to reflect and enhance the lights and visual displays from the discothèque set up upon her luminous hair, and now being used by the girl to constantly glance at her reflection. The next realisation was that we were also positioned in an area where all others in the venue could see us and pay her the attention she craved, to notice her. I was a trophy, nothing more, simply a body sat beside her to establish the fact that she could acquire a boy if she so intended. I could detect no real personality, no affection nor could I honestly claim any sharing of attraction between us. I was simply and utterly just a pet to display upon her arm! My heart broken but my spirit strangely relieved, I made an excuse of purchasing some refreshments, and ran for my life, never to return to her side again. Luckily sometime later I discovered her sister and she was much more fun!

So my little experience of false attraction gives an example of why looks alone are not enough on which to form a lasting or meaningful relationship. First impressions undoubtedly count towards the initial attraction but to obtain the complete package, one must delve deeper. Does having the perfect physical body draw desire from others? Nature tells us this is most likely a good explanation as the women of old, in the days when being a man really meant being the sole provider for the family, would look for a strong back, good stamina and hunting abilities. A wimp unable to hunt successfully or fight off wild animals, protect against raiding parties from other villages or wandering groups of Jehovah's Witnesses would not be considered a good mate. Nor would the unhealthy individual be chosen in times before the blessed state benefits came into being, in history the sick or injured were ignored and sometimes even permanently cured by knife or sword. A healthy physic and good strength were high on a woman's expectation in times past, but what do they expect in modern times? Does a perfect physic still hold such value or has the ability to think outshone the ability to fight? Plainly so in most cases, a husband who fights ends up in prison, a husband who thinks intelligently, gets away with it!

So what does a man look for in his perfect woman? Ignoring the suggestive remakes and insinuating sniggers I will continue. I admit there are most likely men who are only attracted to the purely physical aspects of a woman's body but I suspect these are few, the majority of men are not as shallow or fixated on sex as modern comedy folklore would have us believe. Like the women in our lives, men also look deeper into what would constitute a healthy and long lasting relationship. Okay, I will pause here until the ladies amongst you have finished laughing!

But it is true, if males formed relationships on physical aspects alone, most men would still be single and running bare arsed around the countryside as, without wishing to offend, perfect physic's in women is in the same short supply as in men. So if men sought only perfection, there would be millions of women still available. Again the same would apply to men if women preferred an Adonis like figure alone. But luckily a perfect body is not the sole reason for attraction though it would makes a good first impression. Fine firm stomach muscles are less painful on the eye than a beer belly hanging languidly over a trouser top!

I feel I must add a couple of comments here about how men often describe what they attribute their individual attraction to concerning the female form. How many times has the phrase, "*I am a leg man myself,*" or "*I like a nice bum,*" or less often it seems but still applicable, "*Hey, nice rack!*" Each and every one of us has heard at least one of these phrases at some time, but are these actually true? Do some men form the object of their desire based solely on limbs, breasts or buttocks? Personally I do not think so, I would surmise that along with other aspects of the female form, these simply contribute to the overall picture and cannot be regarded as single examples of physical attraction. I do often ponder what, if any, aspects of the male form would bring forth similar exclamations from a woman as she admired a male passerby. As I am obviously not a member of the female persuasion I cannot answer this, but I wonder all the same.

Are facial features an important part in the initial forming of a relationship? I believe this is probably the case as it could appear under the heading of first impressions. A pretty face, wide bright smile and huge eyes will always draw attention to a woman from a passing male. The face is the first area our attention is taken to, until the males eyes drop to other important features! Male or female, we hardly pause in our stride to pursue further observation of a face that only a mother could love, the face is what gets our attention in the first instance. Once the face and features have been examined, and if they passed our perception of what is deemed attractive, then and only then would we continue our scrutinising of that person. Obviously other features may grab our attention if that person was walking away from us, but this image alone would not normally cause us to catch up with the person and peer at their face. Nope, we simply admire the wiggle as it moves away and continue on our own path.

A pretty face is not easy to define as we all have our own conceptions on what we regard as attractive. Some professor, quango or group of over paid researchers have come up with the notion that we are attracted to those with similar facial structure as ourselves. Huh? Why the heck would I chose someone who looks like me? Even I would walk quickly away from a face such as mine! However there may be a grain of truth in this assumption. A girl with an open friendly face may well attract an open faced man, an obviously intelligent face may appear highly desirable to an academic in the same vain as a thin face appealing to another and a fat face to one of similar proportions, but I believe this may be as far as the trend can go. Although I have mentioned the attraction of a pretty face on numerous occasions, I must add that prettiness in itself may not attract everyone. A plain honest face with little or no makeup, naturally attractive in its simplicity and framed with natural un-styled hair is a huge draw to many men. Likewise I assume the same applies to women who when first noticing a clean shaven, un-adorned male face consider it attractive when compared to the sun tanned, stubble wearing, and slick haired male models that prance across our television screens with an arrogant air.

I realise this is a wild shot, but do pheromones play any part at all in human attraction? As many of us know, pheromones are a major factor in the animal world, from lions on the Serengeti to the randy mongrel next door. Two gentlemen named Peter Karlson and Martin Lüscher coined the phrase pheromone in 1959 and initially identified pheromones as chemical alarm signals. Strange when the word pheromone means *pherein* (to transport) and *hormone* (to stimulate) in the Greek language. But it is in insects we see the most use of pheromones, bees for instance, but many animals certainly do make use of the smelly form of signaling or communication via pheromones. Cats and dogs release pheromone in their urine to announce their presence to other canines and felines. I have no real idea what a pheromone is, be it a smell, a sensation or a chemical trigger? I do not know and have no intention of researching the word any further. I do know it is used by insects and animals to attract or alarm or warn off other members of the same species.

So do human give off a form of pheromone? Can our senses be attracted to another via scent or smell? I can categorically state that a smell, especially that of sweaty socks, bad breath or repugnant body odour will definitely play a significant part in the game of attraction, by repelling all suitors within a ten foot radius! But what if pheromones do contribute to those emotions and standards of attraction? I believe there is no definitive answer to this, but just imagine the possibilities. The person who managed to 'bottle' a scent or perfume that contained a human pheromone for sexual appeal or physical attraction would undoubtedly become a very rich person almost over night. Everyone would clamour to purchase such a product in an attempt to make themselves irresistible to the opposite sex. Put me down for two bottles please!

There is not enough evidence to suggest that humans have a distinctive pheromone production, I simply rambled throughout an odd thought process. However could smell itself help dictate who we may consider to be attractive? Yes I believe this is certainly true. How often does a perfume, aftershave or anti-deodorant linger pleasantly upon our nostrils? A normally plain or unobtrusive woman who may not continuously attract instant attention can alter this situation by administering a scent to her person that accentuates her femininity, boosting her allure to nearby males. I have certainly come across women who have instantly transfixed me with an aroma, a perfume that sets off their other features and personality like no fashion, diet or hairstyle could possibly achieve.

I would assume this applies to men also, a suitable fragrance drifting suitably across the divide as the man and a woman meet in time and space, or in the office, shop or night club. A male wearing one of those over powering, mass produced '*he-man*' fragrances that assaults everyone within the room will often detract from any possible attraction instead of empowering it. We have all been on the receiving end of these too powerful deodorants so we all know which ones I refer too, apart from those young men who consider such an over bearing odour as passionate. So therefore our own individual smells, scent or body odour can and will have a major effect on our ability to attract another. Or repel in many cases if we get the concoction wrong!

Almost all seeking a form of physical or emotional relationship will insist it is the personality that attracts the most, and actually I believe this. I have known many fabulous looking people with the personality of a fish, but also many fish faced individuals with temperaments and personality that simply flows. I do not think we are attracted to others by appearance alone, after all one has to live, talk, laugh and cuddle with the partner of our choice, not so appealing when the other is a complete bore! Personality is the main characteristic that defines our being, our niche in the human mating regime, without this very important feature I suspect we would be `*at it like rabbits*' with anyone and everyone. Not so bad I hear you cry, until you need sympathy or a caring hand on your brow, until you are lonely and require the understanding of a soul mate or until those darkest moments arrive when one really needs loved ones the most.

Personality works in a complete different way to that of a purely physical or first impression attraction. One does not see immediately that another posses a sparkling persona or that of a cardboard cut out, politician or banker. A persons individual character has to be discovered, gently prised out over a period of time and carefully noted against ones own set of criterion. These are often such small things, does he laugh with you but not at you? Does she show how she perceives the world around her or does she expect you to know instinctively? Is he generous or is she demanding? Does he scratch in public or is her makeup over done? Are his shoes clean or is her skirt too short? Is he or she ready to apologise when at fault or do they insist on their innocence and attempt to turn the blame? And of course one of the major personality traits we all seek is that of faithfulness, will they remain at ones side through adversity, through strife, through illness and through those teenage offspring years.

Some consider individuality is not justified as a reason for attraction, instead it is considered that we seek another's persona or qualities which involve those characteristics we ourselves hold as virtues. In this context, people often marry not the person they actually want to, but a person they may aspire to be. Obviously these marriages seldom work as both persons want to be the one individual, two parts of a whole. It would be like marrying oneself, not something I would relish!

All these small questions help form an opinion of another's personality but there are more major characteristics that can result in a mutual attraction. The biggest of all questions that may discern the presence of attraction would, in my opinion be, does he/she make you happy and do you enjoy his/her company? These are the real basis of attraction, one cannot envisage spending a life time with someone who constantly irritates, annoys or infuriates. Let's face it, these minor infringements on a relationship tend to appear after one is married, it is common knowledge once a couple are joined, then the aggravation begins. Maybe for a few weeks and certainly not during the honeymoon one would hope, but they do rear their nasty little head eventually and how we cope with them will place judgement on our conception of attraction, and show an indication of our individual level of personality.

Wealth is an obvious attraction to both male and female in search of the perfect partner, and there are those who actively seek out individuals lucky enough to have wealth or to have obtained fame. Certainly fame is a huge draw to many, consider a Premier league footballer or a young pop idol, how many fans would give their souls to be the object of that individual's affection. The proof of the answer is evident; simply take note of the thousands of adoring fans that accompany these modern celebrities as they clamber from stage screen, to football match, to award ceremonies and expensive night clubs. But can a relationship endure if based solely on fame? What happens when the glamour fades and the once famous pop star develops a large waistline or the international footballer's sparkling career dissipates with age and is forgotten amidst the sporting history of those who were famous once, but where are they now?

Fame alone perhaps would not be the best foundation upon which to base a lifetime relationship, fame is fickle and fades quickly as the population moves forward to adore the next new pop star, the next professional sports person or celebrity. It is unfortunate that in today's media and therefore in the hearts of the populous in general, it is the television, film, sport or music hero that receives acclaim from the masses. When did you ever see a scientist who discovered a major cure for some fatal disease being chased down a street by screaming fans? Would you even consider piling exultations on someone who had simply saved the life of another? No, you may offer a brief word of praise before moving quickly on. Few of us count doctors and surgeons amongst our modern day heroes, few people delight in the achievements of explorers who accomplish great feats, and even less notice the little actions of charity or kindness portrayed by those unspoken and un-noticed real life celebrities amongst us day to day.

Fame cannot really be the sole reason for an attraction between two people, fame is not a construct of personality nor does fame reflect the true nature of that individual. But fame does attract and will always attract those who wish to live their lives through the life of another. On the opposing side of the argument and accepting the attraction of another based purely on fame. Can the person behind the fame be loved for themselves or only for the fame? Perhaps that is why many famous marriages break up, those that survive the test of time often occur when the famous person marries another from their childhood or from the period before they became a celebrity in the public eye. So it would appear that fame rather than the individual is the allure here, possibly a shallow enticement without future.

Wealth itself can be a totally different attraction, not depending on appearance, physic, sporting ability or even age. Most of us have joked at one time or another that we would love to meet and marry a millionaire. A joke or dream to most but many actively seek out these rich people. Is it certainly not uncommon to discover a wealthy person with a partner half their age clinging tightly to their gold watch adorned arm. Obviously we would all like to be rich, to possess the ability to purchase whatever we wish, to go where we wish and to be how we wish to be. Alas this is not the case for the majority of planet Earth residents. But does wealth come in degrees?

Would a man with a bowl full of rice, enough to feed his family for a day be considered successful in his own community? Is an African tribesman with a 100 head of cattle considered as prosperous as a bank manager or top civil servant in Britain? Can a civil servant be considered as loaded as a Hollywood film star? Or can the person who has achieved his life's ambition and now relaxes with enough income for a comfortable retirement be classed as wealthy? Which one is richer?

Of course the main perception of wealth consists of having a personal fortune in the millions of dollars, but in truth, does the actual financial figure matter? It depends on ones own definition of prosperity and sadly today the gulf between the rich and the poor is fast becoming an insurmountable crevasse which few can cross. So is wealth a worthy factor in the ideals of attraction? Yes it is but how would one categorise this form of appeal? There are certainly many descriptions for those people who actively seek out those with large bank accounts, gold diggers for example. However should we criticize these people? After all we would all like to be rich in monetary terms, so does the method behind achieving this result really matter? Can an affluent person ever truly believe their partner holds affection or is it simply self-indulgence? At this point some may say who cares? If a nubile young woman is willing to share ones home, company and bed, does it really matter if that person feels love? Possibly not, many people would give and often do give a fortune just to gain the attentions of the person they desire. If I were a rich man (*there's a song there somewhere*) I may well consider obtaining the personal phone number of Cheryl Cole or Parminder Nagra or even one of those unapproachable girls from my school days. Well possibly I may consider these actions but not actually act on them!

But wealth can provide an easier life in general and who can honestly say that is not the true desire of us all, not having to worry about the electric bill or the water rates or even search the supermarkets for those endless bargains. I know right now that I would prefer to stay in a five star hotel than a small Bed & Breakfast establishment. I have stayed in many B&B's and Guest houses and cannot fault them. It is simply the idea of being pampered and waited on in a huge hotel such as the Ritz that appeals to me. The grandeur, the splendour and the opulence along with total servitude offered to guests by the very best hotels is an adventure in luxury I wish to experience. I suspect this reasoning will appeal to many others also, hence the factor of wealth must be considered in the art of attraction, but however it is not a factor many of us possess so although the dream may be there, the reality is not.

So what are the factors that create an attraction between we humans? Is it the perfect body and or appearance? Certainly this does play a very important part, but is it enough? I do not think so. And do we consider the facial features or 'good looks' of a person a aspect that attracts us the most? Again although this will obviously be a contributing factor, looks alone cannot be considered a form of attraction with lasting results. However I do conclude that in certain situations facial characteristics can be a sincere form of attraction. Possibly a blond and tanned male with a love of surfing and the out door life would be strongly attracted to a blond female surfer rather than a brunette of the librarian style. I realise this is a weak example and I apologise, but I think it serves its purpose in getting my befuddled point across.

What about those tricky pheromones, do they really play any part at all in our attraction to another? No I certainly do not believe this, although a pleasant body odour or perfume can and will often lead to a form of attraction. Emphatically a poor body odour will have the opposite effect, that fact we all know! So it stands to reason that a nice niff will attract while a nasty niff will repel.

Personality I believe is one of the most important factors in the positive attraction between two persons. One rarely falls in love at first sight, this is usually lust and is certainly no basis upon which to initiate a lasting relationship, fun of course but not the ideal start. Meeting a person and getting to know them, learning their habits, their desires and their fears and meeting them with ones own faults is all part of our personalities, how we are seen and perceived by those around us and not by what how we may see ourselves in our own minds. Persona is hardly ever responsible for any initial attraction, though it has been know. Rather it is the hard work of attraction, striving to please and be noticed, unlike certain bodily aspects that command notice immediately. Therefore personality must be concluded as a strong factor in the forming of a lasting attraction by one person to another.

Fame is a factor I am discounting from the realistic requirements of desirability simply because it cannot apply in the vast majority of personal attractions rampant in the world today. Yes one may have strong desires towards a famous person, but can one achieve a response to this, I think not. So fame although a definite lure is simply that, a lure!

Wealth on the other hand is a factor that many people seek, either by ensuring they are always in the vicinity of those considered rich, by offering certain favours that may appeal to the more mature wealthy person or simply by offering a form of attraction that the well-heeled person desires. Look around at any Formula One race track, night clubs where footballers frequent, stage doors or wealth saturated cities such as Monaco or Los Vegas. Here one will find intermixed with the indigenous population, huge flocks of beautiful bevies and tanned hunks all seeking that escape from poverty or normality by ensnaring an unsuspecting individual with a bulging bank account. Be honest, have you ever seen a rich man with an unattractive woman hanging from his arm? This example is less frequent in rich and powerful women but it still does occur. I believe the Americans have a word for this type of wealthy single woman of mature status, they are called Cougars. Why I have no idea but there it is, a rich man with a young beautiful female attached like a leech is considered either lucky, or a dirty old man. A woman of sound financial stock with a toy boy in tow is simply called a cougar. The logic of justice appears to slightly array here but no matter, in the case of attraction by wealth, money is the over-ridding compulsion. Wrinkles play no part in these relationships.

While pondering the subject of wealth, another thought crept into my mind, what if both parties were each as financially steadfast as each other? Does money attract money? Of course it does. But what of 'old' money? That ancient wealth that identifies the aristocrats, royalty, the so called 'upper classes' and families with a long reaching grip on history? Could an inherited fortune trigger the response of attraction to another of the same distinction? I conclude it does, and that is where the historical system of breeding came into place, creating the class system that no country can really escape or deny. But can an arranged marriage staged simply for status or social class be attractive? Could a Prince marry a pauper? Would he even want to?

So I conclude that breeding or class would not, in most cases, feature as a factor of attraction, not in the real sense of the word. Though many young girls still dream of their Prince on a white horse, young boys may still occasionally dream of a Princess in a white Ferrari, but they are still just dreams.

Attraction is seems is not an emotion that relies on one factor alone. There are numerous factors involved, many more than I have discussed here of course. But I conclude that looks, physic, humour, persona and the ability to provide financially rate as the most likely aspects that contribute towards an attraction between two people that may stand the test of time.

Chapter Seven. Class and Breeding

Just for a change I am writing this in daylight instead of the dead of night when I seem to do most of my work. Perhaps this section will make more sense as I am wide awake and able to concentrate, however I would not expect too much if I were you. My insomnia appears less frequent these days, though I admit it has been some time since I began work on this project. So now instead of the delicious drink of hot chocolate, I am now sustained by a faithful old cup of tea. Anyway I appear to be meandering again, so back to the grindstone.

Following on from the previous chapter, my strangely awake mind began rummaging through the distinctions of social class and historical or possibly hysterical breeding. I have already concluded that neither class nor breeding can be considered a major factor in the process of attraction, but this leads on to the subject of class itself. Does the status of social class still exist, not only in Britain but also around much of the world? In many areas of the world, class is difficult to define but it is there without doubt. Class is not solely related to money, it is also ones breeding which could explain why so many wealthy people of today have no idea how to portray themselves in public. The antics of over paid footballers, minor pop stars and wannabe celebrities prove that money does not buy intelligence, common sense or grow much needed brain cells.

Class can be related to income in certain circumstances, but upper class people are still those who have education, considerable wealth, family background, and connections with those in positions of power. Celebrities and the nouveau-riche have money and visibility, but are lacking in the other areas that make them part of the upper class. A large element of being considered upper class is the cultural behaviour that helps a person function within that group. Possibly class distinction is slowly fading away as working class and middle class are now hard to distinguish from one another e.g. a working class person could have two cars, as could a middle class person. But what has long been established as the true definition of class relates to ancient bloodlines that can be traced back centuries. It is not money. There are many multi-millionaires who are far wealthier than those with the legitimate title of upper class but they do not have the breeding, the history or the social standing and so will be seen for an eternity as only affluent middle class. Wealthy and rich they may be but they cannot claim to be a true aristocrat. Can you imagine, those of you who happily possess more than one brain cell, Prince Elton John or King Paul McCartney, or God forbid, Queen Madonna!

I have been pondering this line of thought for many years now and still I have not arrived at an answer, the thought is, does class distinction actually still exist in our modern society? Many people who may consider themselves modern and well educated have suggested that a class distinction no longer exists, deeming it buried in the dark ages. I do not believe this and one only has to watch television news or open a newspaper to hear and read about royalty, not only in this country but others such as Spain. With royalty there must come a considerable lineage of inherited nobility, ancient families and wealth accrued over centuries. Perhaps the self styled modern man is in fact the court jester, failing to notice the wider world around him, instead keeping his vision and understanding firmly locked in his own back yard, so to speak. So yes class does continue even in modern times, but is the distinction between class, breeding and social standing still as highly prized as ever?

I realise that the lower classes no longer doff their caps to the good gentlemen of the manor, nor do housekeepers, cleaners or cooks have to bow and scrape to their Lords and Ladies. Parlour maids and kitchen girls no longer suffer the attentions of the red faced, over bearing Baron or the head of a large family estate. Stable lads can in this modern age, bend over in safety as they perform their numerous duties without the fear of a rear assault by an aged and confused Squire. Of course I am exaggerating conditions of employment between an aristocrat and his/her servants but in truth for many centuries the wealthy, the estate owners, the Lords and Ladies of the Manor, those related to royalty and royalty itself held huge power over those beneath them on the lowest rungs of the social ladder of humanity. So though I jest, I do so in order to set the gulf of difference between the two distant poles of society in history.

In centuries past, class played a huge part in society. In the British navy for instance, an inexperienced boy of 12 years with a rank of Midshipman could demand respect and obedience from seamen five times his age and with a life times experience in the ways of the sea. Children of wealthy parents regarded as middle to high class had a frightening amount of power of those considered beneath them, while adults of class could often hold the rights of life or death for those unfortunates amongst the lower class. Right up to the early 20th century, class held sway over the lives of all populations around the civilised world.

Today, in the western world, no one has to bow or grovel to another, simply because that person may have more money or wealth, I will admit here that I have met one or two that still seem to expect servitude from those less fortunate than themselves, obviously they frequently fail! But are we still bound in some ways to the class distinction of old? Do we still hold those with wealth or lineage on a pedestal above us, fawning over and idolising them? There are many who would say yes, many others who may not admit it but still do, and of course many who would answer a firm no to this statement. Some families retain strong links with their traditions of wealth marrying wealth and breeding wealth, so there is an undercurrent in society marriages that still meanders on. However people today are free to take their own road so it could be seen as an improvement to the non-aristocrats among us for the even distribution of wealth is a definite benefit for all.

When we examine the subject we can not escape the true fact that most of us do consider someone or some persons to be on a higher plane, a higher social standing or higher class than ourselves. Today it may not be purely down to wealth or breeding, it could be a football player, a pop star or actor, and it may even be a politician though that is most unlikely, but you never know. . . . . And that certainly has little to do with breeding!

Is the position of idol the same as high class? Or does the notion of class still depend on the amount of money, breeding and heritage a person has? Perhaps in this age of entertainment the two go together, fame accompanied by fortune, footballers being paid huge amounts for kicking a ball round a park, or a singer receiving millions for singing a song or an actor pocketing a fortune for pretending to be someone else on screen. Could The Beatles be considered upper class, instead of four lads from Liverpool? Has their fame and obvious fortune risen them to the greater height in the social class distinction, or could their social standing be accredited to another reason such as wealth? Of course it must, everyone knows the origins of those four lads, class had absolutely nothing to do with how people became to idolise them. Fame cannot be considered as a class distinction, no matter how much wealth and social standing that individual may have. No one could ever describe the four Beatles as being 'upper class', even those of the band that are still with us would, I am sure, deny accepting a class status they felt inappropriate to their upbringing and origins in the city of Liverpool.

   I realise that of old, class was almost entirely related to inherited income, but today those considered upper class are mainly those who have the best education, considerable wealth, family or historical background and connections with those in positions of power. People today, least in the major countries may not consider aristocrats as an ideal biological factor in their own choice of breeding partner, but certainly many still care deeply about a bloodline when considering a mating partner. This factor of course cannot completely override emotion, but many do care whether their potential future mate has different genetic traits, such as predispositions to illness or mental disorders. With our modern scientific knowledge and understanding of inherent diseases, deficiencies or defective genes, many now carefully consider their possible partners genetic history. To scores of us it is only natural to care about the genetics and heritage of someone with whom you may have children. It is a survival instinct and certainly has nothing to do with someone being wealthy or not. But of course, genetic breeding for health or appearance has little to do with class, a fact our old mate Hitler should have considered before attempting to breed a superior race!

Maybe we are on route for a simpler two class society, evolving from the three classes of past centuries, possibly the structure for middle class has been lost, a huge void now between upper and lower classes.

So where do we place the `*new rich*', the footballs, celebrities, lottery winners, pop stars, movies stars, oil barons, media moguls, self made business millionaires? What class would be appropriate for those with wealth but no social background in society or history? Should there be a suitable social class status for those who are pig ignorant but have managed to obtain huge wealth? Yes I realise many will be able to identify such persons; they frequent the tabloid newspapers more often than publicity seeking politicians. Being upper class is often considered to be due to one's social behaviour and practices. Perhaps it is the higher state of mind or behaviour or even confidence built up over centuries of controlling others that raises one up in society. If you feel superior then you will become superior. Least that is what my psychiatrist told me . . . Class I feel cannot be considered linked to ones personal wealth, class is a state of mind, a personality and an inherited standard of behaviour and accountability. The stiff upper lip and straight back bone combined with the abject fear of offending ones peers all contribute to what we consider as social class. Grace and perfect manners, bearing and pose, knowing ones position in society and feigning a total devoid of emotion is compulsory in the regard of class distinction. A bright yellow 'shell' suit with bulky gold bling and a vocabulary that consists of only four letter words cannot in any frame of mind be mistaken for aristocracy! Nor can the simply ability to talk in circles but actually say nothing be considered as class, nevertheless this trend does appear to be growing amongst the financial Mongols, the established politicians, the hierarchy of media and the music industry. How ever one attempts to portray a high degree of social class, it is not a title that can be sought and bought like a knighthood or peerage. Let's be honest here, how many of the members that grace or disgrace the House of Lords can lay true claim to heritage and class? How many members are actually Lords and Ladies via an inherited lineage? Nope, I am sorry but I can not come up with a single name. My ignorance I admit, but hey, who really knows and more to the point, who cares?

Although I am not against the House of Lords I do resent the fact that many top civil servants and Members of Parliament all expect a peerage upon retirement. When one is aware of those who spend their lives searching for a cure for cancer, those who have given over their entire working lives to helping others and those that have made major improvements to the lives of other less fortunate persons receive nothing. Not a peerage, knighthood or slap on the back!

Footballers, sport stars, pop stars and actors regularly receive awards, Sir Cliff Richard and Sir Elton John for instance, but a lowly scientist who through hard work and perseverance, finally develops a new drug to combat a serious disease gets a state pension and a bus pass.

But these Lords, Ladies, Barons and Baronesses are, for want of a better term, manufactured by modern day principles, the true aristocracy are still defined as those with a family history going back centuries, wealth passed from father to son down through the lineage along with certain expectations of behaviour and social position. So is the concept of social class only the result of breeding or heritage? If your father is a Lord, can you expect to be elevated to a class above others whose father's did not have title? Is this still relevant in our new and modern world, do we still care what a persons title is, be it Lord, Sir, Lady or Majesty? Well maybe Your Majesty does still relate to class, one would hardly address the Queen of England as `Mrs Windsor'! What about other titles, doctor, professor, managing director, do these titles convey the same level of importance? Other titles to consider may be Prime Minister, President, Chancellor, Chairman or Emperor, does one's class play a part in obtaining these titles? No of course not, all these titles are political and may be obtained through a political or military career, or bribery and corruption depending on what country one chooses to rule. The ultimate question here would be, could a lowly corporal who obtained the high title of Chancellor of his country be considered as a true representative of the higher class society? Baring in mind the history of our planet, perhaps not! But can one be considered as a member of the class society without a peerage, heritage or royal bloodline? Perhaps, but only if one has the bearing, pose and attitude to carry the deception. I have in mind a television presenter of antique shows. Always immaculately dressed, possessing a high intelligence and with a deep knowledge and understanding of history, heritage and culture.

Obviously simply being a well dressed and well spoken gentleman of the television itself cannot justify the title of class, but rather a portrait of how one would expect an aristocrat to appear. Possibly others may also fall into this bracket of being considered highly in social standing without the background one normally associates with class.

Does the pursuit of class reflect on the ordinary person's expectations in life? Yes I believe it does to some small degree. We are all encouraged to marry or seek a partner from a higher social status, for instance how many young girls are cajoled into seeking out a prospective young doctor or lawyer as husband material? How many young starlets are, for the sake of their individual career, pushed towards an established media personality? What about the 'arranged' marriages in Asian countries, girls and boys forced into marriages deemed suitable for the families involved. I am not considering the prospect of a princess seeking only a prince for her union, but matrimony for the sake of profit, either a gain in social standing or financial advancement do still play a major part in our lives and can be seen as a form of class breeding. Lately our own United Kingdom royalty has made a conscious effort to marry outside the usual gene pool of hereditary aristocrats. So does the term 'class' still carry the same implications of years ago? Or do we now see class as a reflection of ones own abilities, social graces and status?

As far as breeding goes, I am not quite sure what benefits there are to being able to trace ones family history back through time, I fail to see how this can assist in modern living. Conversely I conclude that the world will become a better place once racial, class, and religious distinctions are ignored, or at least merged together with all the races combined and so race nor class or breeding are no longer distinguishable, I believe there will be less discrimination and the world will be a better place to live in!

So does class distinction actually still exist in our modern society? Yes I conclude that it does, though in a far more diluted form than over past centuries. Class can be defined as family history, wealth and power. A distinction that has continued over time with aristocrats and royalty continuing a bloodline throughout the eras and establishing criterion and bearing that often stands prominent amongst lower classes or castes. Class cannot solely be attributed to countries such as Britain, Spain or Sweden alone. There are now only ten monarchies left in Europe so possibly this is a sign of which route the nobility find themselves on. At most points in history every country had a King or Queen, but now it appears this trend is on the decrease. With no monarchy, can there be Lords, Ladies or other such heritage positions of class structure? No of course not. Without a monarchy there would be no logical reason to retain titles relating to or stemming from connections with royalty or civic awards such as Knighthoods, Barons or Dukes. Titles would disappear and along with them the ancient status symbol of heritage and royalty. Politicians, business magnates and megalomaniacs would rush to fill these gaps in our society as swiftly as the aftermath of a strong curry can clear a packed room! No breeding or history, no conception of standards and no loyalty to those classed as below them on the social ladder. The elite in our society would be degraded from families with history, heritage and social class down to those status in life exists only on multiple figures on a bank account.

Class distinction does still exist in many parts of the world, not only in Europe. Many countries such as South Africa, Uganda, the Americas, Asia and Malaysia still retain a royal family and one would assume all the relatives, family friends and assorted hangers on may still be considered as upper class. I suspect in one or two of the countries mentioned, class continues to be a major factor in society and one may still lose ones head if one upsets a prominent member of certain royal families.

In the western countries class is fast becoming indistinguishable from wealth, millionaires abound and buy up ancient estates and manors in order to promote their own success. With the Lottery a road sweeper can achieve wealth over night and entitle him to rub shoulders with both the new money and the old. A winner of one of those ludicrous television talentless shows can wake up one morning to a new status of wealth and celebrity without a single notion on how to deal with the situation. A successful businessman can achieve a rank of nobility by simply becoming rich, not unlike a certain Lord with a very sweet sounding name . . .

Breeding however does continue, old money and established families still seek marriage and reproduction within similar circles, obviously with a few exceptions such as Grace Kelly, an American actress who married Rainier lll, the Prince of Monaco. But in the main it appears that those who have a long family lineage still prefer to marry and breed within their social status. But perhaps this is also a trend that may be showing signs of decline, especially with the recent marriage of Prince William and Kate Middleton. While this is acceptable in our present times, most with a history of breeding do appear to adhere to the principle of marriage within their own section of the aristocratic community.

So the system of class and its hold on world dominance may possibly be waning, diluted by wealth from other sources and corrupted by the 'instant' rich, media personalities, successful sportsman and business magnates. Class will always exist in some form or another as long as there are Kings and Queens reigning over us. Breeding however continues its path through history. Though little mention is made these days, ones breeding is still considered by many to be the ultimate factor for consideration when selecting a life partner. Strained and diluted by wealth and fame, our modern conception of social class and breeding, are two qualities within society that march on.

We are all aware of how class and breeding play a part in our society, but how do we consider those newly important super beings simply known as authority? Those mysterious persons proclaimed by all as *they*!

Chapter Eight. They!

Some things change but my sleep pattern does not, damnit! Once more I am wide awake late into the night as sleep eludes me, so why can *they* not do something about it? The rest of the house is sleeping soundly, if not quietly but at least the sound of snoring assures me all is well. *They!* That is a very big word in today's modern world. *They*, it is a word we all use frequently throughout the day, a word that in its very meaning encompasses entire sections of the human race, so who are *they*? Speak to anyone in the street, in the home, in the office or even in a shop and sooner or later the word *They* will pop up in conversation. Maybe it is more of a British thing rather than world wide, perhaps it is just we English who insist on blaming someone else, blaming the elusive *they*.

So who are *they*? Are *they* the same as *somebody*? Could the word simply refer to a Council, authority or the Government? Is it a word that has become associated with authority? Strangely we do frequently refer to those in charge or those in authority as *they* or *them*. And when speaking of authority it will often include County Councils, Parish councils, medical establishments, large businesses and other assorted organisations that endure the brunt of any initial complaint in society, after the Government that is, we do have our priorities! But as an example, if a person were to trip over a raised paving slab in the street, I guarantee either that person or another will ask:

'Why do not *they* do something about that?'

'It is about time *they* fixed that.'

If a favourite landmark was to be demolished voices would ring out loudly, along with waved banners and misspelt placards.

'Why are *they* allowing that?'

When our taxes increase, a multitude of tongues, rich and poor will proclaim with indignity across the land.

'*They* should not be allowed to do that.'

'What are *they* doing with our money?'

When people talk about authority, the word chosen as a label is *they*, for instance if a young man cannot find a job someone will say; '*They* should help.'

When a drain breaks in the road, in will be '*They* should fix that.'

Other statements are also frequently uttered like,

'How can *they* do that?'

Or '*They* should not be allowed to get away with that,'

'*They* didn't say,'

'*They* do not know.'

When talking about the medical profession for instance. Rather than simply name the hospital or doctor or medical department, surgery staff or even just the family doctor, people always seem to revert to the word *they*. For instance,

'So what happened at your appointment this morning?'

'*They* do not know what's wrong with me; *they* think I may need a brain transplant or something.'

So why do we use the word *they* so frequently? Are we just lazy and cannot be bothered to give a full title? Maybe it is habitually the fact that we just do not know who *they* are in certain circumstances. Conceivably it has become enjoined into our way of life, our verbal antennary, and our method of communication. Mayhap it indicates our lack of deep thought in such a way that we no longer see any problem with entitling all forms of authority as *they*? But perchance it is a simple degradation of our day to day language, the need to get a point across in the quickest possible way without resorting to full titles or explanations. Although in our modern country all persons are acclaimed to have a basic education, a clear conception of the English language and certainly a wide range of assorted communication aids, this is sadly not the case.

From experience I know there are a high percentage of adults in this country who have difficulty in reading and writing, mostly through no fault of their own. Giving some credit to the fact that some 60 million persons now inhabit the British Isles, it is not surprising that some unfortunates may slip through the educational net. Others may have had unpleasant experiences at school, dyslexia is on the increase or possibly just our awareness of it, and for some the academic process is too long, resulting in many young people leaving education for the work market. Unfortunately a minority consider enough is known already, mostly at the ripe old age of thirteen, and cannot see the point in continuing education when they are so brilliant now. So does a lack of education explain our constant use of the word, *they*? I do not think so.

Finance is important in how one is educated, and in truth most residing in the western hemisphere have the opportunity to gain a suitable academic education. Finance may and most likely does play a huge part in how we all see our individual direction of learning. A family on a low income may wish to see sons and daughters out earning money and helping to pay their way to reduce the financial burden on the family as a whole, rather than staying in the academic system. For those young people with lower educational expectations, academic expectations are the point being made here, not academic abilities. Those students who may have no interest in learning about our past Kings and Queens, about the geography of the world, about how atoms and particles react when mixed with another. I can understand them not wishing to learn about past Kings and Queens, how is this information going to assist ones efforts to find a job? I always hated history, I remember a Wurzels song I think, entitled *"I weren't there so I don't care, so don't tell I, tell ee!"* Sadly my history teacher did not share my point of view, gaining me a slap with a ruler across my knuckles for stating such heresy! Look at it this way, is telling a potential employer that one knows who reigned over our country in the eighteenth century really going to impress the heck out of a garage owner or supermarket manager! Knowing which end of a screwdriver is the handle and where to buy stripped paint may be a bigger advantage.

It is a fact that a practical minded youngster may not wish to fill his/her head with random facts, dates and figures, when all these individuals really wish to do is build, fix or manufacture physical items. Items such as vehicles, houses, machines or create life through growing food, possibly even gardening. Art also does not require a qualification in Astrophysics or accountancy, Art requires a passion, the ability to create, or as a way out of hard work! Not everyone is suited to academic study and this is a very good point, otherwise who would build our houses and fix our cars, a bank manager?

So why do we insist on pushing these people into a classroom and attempting to fill their heads with information that will have absolutely no bearing on their future? A good academic education is not the only way to succeed in this modern world. People with little or no classroom interest may turn out to be the mechanic you take your broken car to, the person who builds your house, or removes your rubbish, or drives the train you desperately need to get you to work on time. Not everyone chooses an academic route to a career, but without the skill input of the vocational student or apprentice, our economic, political and logistic superstructure would very quickly collapse. No farmers, no firemen, no policemen, no taxis, no plumbers, no electricians, no sailors, soldiers or airmen, no road sweepers, no roofers, no gardeners, no waiters or waitresses, in fact if you look at it this way, an academic education is not that vital to the way we live after all. Yes I accept we also need those who did decide to pursue an academic route in education, the doctors, dentists, lawyers, scientists, nurses, teachers, lecturers, politicians - yes well maybe we do need the odd politician though we may hate to admit it. Wait a minute, did I say lawyers, what was I thinking! While it is a fact that we need the academic or professional individual, it must also be recognised that we also depend a great deal on the practically skilled member of society.

Education or lack of it still cannot be held solely responsible for our falling communication skills and hence the increased use of the word *they*. Television and home computers, emails, mobile phones and text messaging, popular films and even class differences and expectations have all begun to erode our basic skills of conversation. I mentioned class differences here, but another word if I could think of it may be more appropriate. I am constantly being accused of being a *snob* due solely to my efforts to communicate in a grammatically correct manner. I have been described as a *flowery* talker, a *fancy* talker or even posh. Hah! I am actually none of these; I only wish to communicate to others in a clear and concise manner, with a reduced possibility of being misunderstood. I know from experience if I let my mouth speak without first engaging my brain fully, the results tend to be a cross between utter rubbish and incomprehensible waffle!

It frequently appears that some sections of our community take pleasure or satisfaction in dumbing down speech, knowing the correct grammar for the situation but refusing to use it. Why can we just not say what we mean? Why do we insist on avoiding verbal labelling of titles we all know or should know? Have you ever tried to understand and decipher a teenager's text message? It is almost beyond comprehension, but the young person receiving that text message understands it fully. Strangely enough, this form of language appears to have caught on across the age gap, even I have been known to shorten words or use acronyms when texting, though I do have to consider what I am trying to say. Some of the acronyms used today are confusing and it is easy to mean one thing but actually say another. In fact I wonder why local educational authorities spend so much time and effort in long courses teaching the secretarial shorthand script when most young people are fully converse in one of their own.

With the onslaught of electrical communication, many of our own verbal communication skills appear to be declining. Are we now seeing the initial birth of a new language, a new abbreviated form of speech, of communication, of verbal contact? Have we forgotten much of what is commonly recognised as a universal language, English? Or is the initial statement correct and we are simply getting lazy in our speech? Perhaps none of these questions are true nor a correct reflection of the underlying causes. Is it possible that the frequent use of the word '*they*' is just a defence mechanism? By not actually stating the title or name of who we are blaming, are we not avoiding taking responsibility for our statements?

I am not belittling the use of the word *they*, nor am I associating it with a poor education, lack of understanding of authoritative bodies or medical professionals. No I think the increased usage of the term they is solely down to how our language has evolved. Even if some of its users are still someway down the evolution ladder.

We all commonly use *they* as a general description, it is far easier to say *they* than, for instance, the local council planning department or Her Majesties Revenue and Customs, who by the way often get called much worse! So it appears there is nothing wrong with the constant use of the word *they*, it is in fact just me, I am in the minority of one and should stop rambling on about one single word, before *they* come and take me away . . .

Now that contemplation is out of the way, maybe I will listen to some music, to clear the confusion in my mind and the ache in my brain. Music in its many forms can be a perfect form of communication and a considerable form of release for those who attempt to think too much.

Chapter Nine: Music.

My mind is totally buzzing after contemplating the question of social class and breeding. Also I am finding it slightly strange sitting here in broad daylight as I attempt to put my thoughts onto paper. It is pouring with rain outside and the weather man has cheerfully informed we listeners that a month's rain is expected to fall over the next couple of days. Great! Where's me wellies? Or perhaps a wet suit and flippers would be more appropriate. Last nights sleep has done little to refresh my mind or body, because in truth I did not sleep much. Not due to the insomnia this time, but simply the result of a stuffy and humid summer night. Yes one can tell it is a British summer because as I have already mentioned, it is raining!

During the very early hours of the morning I lay there on my bed in the darkness while those around be slumbered, blissfully unaware of my minor plight. I decided against getting up from my bed and trying to write as I did not have any idea of what to say, so I did not. Eventually my thoughts released themselves from the mayhem of what *they* should be doing and other such frivolities and I began to consider the subject of music. I had no choice really as this seemed to be the path my mind had decided to tread. Strange I know, but there in the darkness of my room and sweltering in the humidity of a wet June night, my mind strayed wistfully and somewhat annoyingly to a tune I had heard earlier that day. Before I knew it, the damn thing was whirling round my brain like a moth round a light, and it would not go away. Finally I gave in and allowed the song to play itself through but I was careful not to burst into song at that hour, a sharp elbow in the ribs administered by an angrily awoken spouse can do serious damage to an unsuspecting rib cage! Besides, humming the tune inside my head seemed preferable to the sheep I had previously attempted to count, the damn things must have been on steroids because they moved so fast I could not keep up and kept losing count! So I settled for the little annoying song, however as my mind was obviously wide awake it was not long before I began to ponder what we define as music in our modern society. Should have stayed with the sheep!

At last the alarm clock sounded its strangled chicken noise and signalled the start of a new day. It was with some relief that I began the morning chores, such as avoiding those in the house to whom sleep had been heavy, and if annoyed in that state of reluctant awakening, could become extremely verbally abusive to anyone showing even the remotest sign of being cheerful!

Tea, toast, and paracetamol consumed by the bottle and instructions regarding the list of chores for me was up to date and suitably long, I found myself finally alone in the house once again. Therefore I returned to my contemplation of music, its definition and the huge assortment of styles, beats and genres that happily assault our tender ears throughout the day.

With the technology available to every one today, music has become a familiar background to our lives. From breakfast tables to cow milking sheds, from the teenage bedroom to the mechanics garage, and from the personal IPod to the car radio, from a myriad of electronic sources music pours into our lives. Music in a wide variety of forms, some music is immediately recognisable while other melodies may fall strangely upon ones ear. It seems everything requires the accompaniment of music these days, even the news programme I am watching at this moment is using a background melody as it announces the headlines. Like the awful canned laughter used in comedy shows to cover the fact that the programme or comedian is not actually funny, music is used relentlessly in entertainment both on and off the assorted formats of media.

Music invades our every moment of existence, some welcome, some not. Blaring beat music piercing the blanket of night as party revellers declare their inconsideration of neighbours and other residents as they drink themselves senseless, or in most cases, drink in an attempt to appear popular even if it is only in their own minds. Music propels us as we shop, supermarkets have long been aware that customers shop faster to the background of up beat tunes, slowing down and examining more goods to the gentler and softer strains of music drifting throughout the shelves. Music with a quick tempo is used during peak shopping periods to hurry the unsuspecting customer through the store, a slower and steadier tempo helps decelerate customers down during quiet periods in an attempt to ensure each customer browses in full the items upon the stocked aisles. One must wonder how fast a pensioner would zoom about the aisles accompanied by the rapid tempo of Sabre Dance by Love Sculpture!

Music provides the back tracks for the vast majority of media adverts, the best and most catchy of which can linger in the brain causing the victim to hum the tune endlessly throughout the day and annoying all within range of this audio torture, or linger like a persistent toothache in one's head. Just like that little tune that invaded my brain in those lonely hours of darkness.

Music or an irritating rearrangement of a well known melody is frequently played in lifts or elevators as one may call them. This is laughing termed *musac* and can result in a real phobia of lifts developing in the poor traveller fearing for their sanity. Music is now regularly played on public transport, buses and trains being at the foremost, in a vain attempt perhaps to settle the agitated passenger and distract them from the fact that their chosen form of transportation is dirty, uncomfortable and as usual, running late! Doctors' surgeries, dental waiting rooms and hair dressing salons all insist on music in some variety being forced upon the sensitive ears of those sat patiently on hard chairs as they await their turn for medical meddling, dental interference or hair customisation. Music has made its way into almost all venues of public attendance, one wonders how long before the Prime Ministers Question time is set to a rousing crescendo of classical music that erupts whenever a politician makes a sensible statement. Actually thinking upon that event, I feel we should be quite safe as sensible statements in Parliament are few and very far between! Even now as I watch the Queen's club tennis on the BBC which followed the news up date, I keep expecting some dramatic music to burst into sound when a player battles towards a match point!

But the most recent modern infringement upon our hearing is that of the mobile or cell phone. No matter where one may be, in a park or restaurant, in hospital, standing in a queue at a theatre, or begging on ones knees in front of the Taxman, it can be guaranteed that someone's phone will suddenly begin emitting a tinny and irritating imitation of a popular song. Ignoring all who turn to identify the offender, the phone is invariably answered with the individual assuming the caller is deaf as they shout into the communication device, perhaps to ensure everyone in the vicinity is aware they have received an important message that simply must be attended to. Sadly in the case of mobile phones, it is not the initial surprise and annoyance of the phone tone that distracts those nearby, it is the deafening volume of the person answering! And speaking of annoying distractions, the car stereo, radio or CD player is rapidly becoming a popular reason for technical suppositories! Young men and women driving around their neighbourhoods with a constant bass beat throbbing from their vehicles. Day and night these future hearing aid wearers shatter the bird song of the day or the silence of the night while attempting to entice a member of the opposite sex into their private little mobile dens of iniquity!

Finally of course there is the abundance of radios, stereos, CD players, IPods, television and computers that blare out music constantly from almost each and every home. Be it so called popular music, rap, soul, R&B, rock n roll or classical in the technology rich countries, music is a huge part of our lives. We all have our own taste in music, some love music as a form of art and appreciate it in all its forms. Others declare themselves as musical connoisseurs when actually they are just following the herd in their choices of music fashion at that time.

The structure or style of music we each enjoy largely depends on several factors, our place in society, our place in a community, our age, when we were born and it would appear, our intelligence! In general our choice of music reflects our age, we all know this, but one can actually give a simple explanation to our taste in music, though I have no doubt there are exceptions, somewhere, somebody will disagree. But not to worry, they will be in the minority as this connection between age and musical taste is readily accepted. So when were you born? Was it roaring forties or the fabulous fifties, the swinging sixties, glamorous seventies, techno eighties or the nondescript nineties? Which ever decade one was born does not really matter. It is the decade when one achieved the mid teens to mid twenties that appears to have the strongest bearing on the type of music we enjoy. So a fifties child will most likely prefer the seventies glamour while an eighties child may well enjoy the most recent musical trends. It is certainly not unknown for people to fall outside this criterion however in the main, this explanation proves true.

Obviously other factors influence what people listen to or play as musicians, a good number even have minds of their own and decide for themselves what style of music they prefer to have assault their ears. However, excluding the real tone deaf persons, those who proclaim to follow only the utmost modern music, tend not to be music lovers in reality but celebrity followers. In wealthy and celebrity over inflated countries, the music style favoured by many will follow what is pumped out by the music media and music stations on the radio or television. Certain television personalities also have a huge effect on what artist's people follow and which trend of music they purchase. In this case it is the new comer to the music scene, the talent show winner or the boy with the cutest looks who is excruciatingly promoted as the music media's latest hero, and followed by devoted young fans that seek out every form of information relating to their singing star. Until the next year and the next young star appears.

This style of music tends to be highly forgettable and is produced for the most rapid response in sales and profit by pampering to those who enjoy following trends or by seeking to inflame the innocent imagination of the inexperienced youngster. Constantly the larger record, CD and DVD stores assist in this subliminal brainwashing of those members of society who cannot seriously be termed as music lovers, the celebrity followers. If one should ask the average young person who is the greatest singer in their experience, the vast majority will chose a star that is in the music charts or being actively and aggressively promoted by the musical media hierarchy at that very moment in time. No thought or consideration is given to last years favourite talentless show winner or chart topper. It is similar with television personalities, famous actors are pushed aside in preference of the young hero appearing in the latest film or television programme. But it is not only young people who show fickle appreciation of music or film celebrities, many older, more middle aged people also fall into this trap as they desperately attempt to convince themselves and others that they still have a finger on the pulse of modern music and entertainment. It can be rather sad to see a forty year old woman fawning over the most recent seventeen year old singing star, while their own children cringe in embarrassment. Or a middle aged man dressing as a Goth and hanging around music stores in an obvious ploy to appear younger. Either that or he's a dirty old man and should be carted off somewhere quickly!

Luckily not all follow like sheep in their appreciation of good music. Ignoring the commercial trends, a vast number of music lovers choose their own style of music, often based on their social group, educational peers and life style. For example a heavy metal rocker is seldom seen amongst fanatical manufactured boy band fans, nor can one expect to discover a spiky pink haired punk lover lurking amongst the hordes of screaming fans of the latest talent show winner. This is certainly not set in stone or The Rolling Stones even! If the song is generally good on its own merit, it will be accepted across the differing dimensions of musical appreciation. But in the main, a person is more likely to follow the style of music that reflects their social position and the decade when music first began to influence them.

The biggest critics of music are of course the musicians themselves. No matter what style or genre they follow, a musician will always notice the melody and composition first with little regard for how short the female singer's skirt is or how many cubic feet of mousse is plastered on the hair of the pretty young male singer. It is the music that counts and stars such as Michael Jackson have achieved a following amongst heavy metal musicians, punk rockers and even skin heads from the seventies, though most would be very careful to whom they admitted this fact!

Too many music stars of today have little understanding or background in music. They simply belt out a song into a microphone, normally in the exact same style as a popular fellow artist topping the charts at that time, but have no idea what musical key they are singing in, nor the tempo or even the instruments used in the composition of their song. To them the music itself magically appears from inside a sound booth or from a computer. Luckily these performers are few when compared to the vast music machine chugging away right across the world. But when such a performer does hit the big time, often the fall is bigger. It is these poor clueless wannabe's that are liable to become victims of the musical media, suddenly an over night sensation with a number one record, but just as suddenly they disappear into obscurity.

So how can we define the music media and the music industry of today? Are there any involved in the actual entertainment business that can honestly claim to be a music expert, or has music become simply another commodity to be advertised, promoted, sold and profits acquired. The music industry itself like any other industry is controlled by managers, shareholders, director, accountants and of course the obligatory bank. The old belief that music companies would encourage and protect their talented musicians has long gone. Now we have corporate millionaires with dollar signs glowing like the mark of Satan in their eyes, not caring about the progression of music or the careers of their protégées, only money. And when we examine the singers and musicians themselves, is there any future for them anyway? The era of vinyl when artists were required to sell hundreds of thousands of records simply to get a toe on the chart ladder has passed. Downloads, CD's and the technology available to all has had a dramatic effect on how we purchase or obtain our music today. Ordinary people have access to and the knowledge to use home recording computer programmes that allow a talented individual to produce an album in their bedrooms and promote it via YouTube.

The music industry has to work much harder to achieve a profit today than in the 50's and 60's, possibly including the 70's also. So it may be good business for an agent or promoter to manufacture a musical artist and ensure they fit into modern expectations in taste, design and of course, star quality. We the public have little say in what is promoted on radios and music television programmes, apart from casting our vote for our choice of new talent on one of the variety of televised talent shows swamping our screens today. One only has to listen to the top forty chart tunes at any one moment to identify the similarities between them all, originality being ignored in favour of instant fame and profit. However it does appear that the public are beginning to suffer from an over dose of cute little boys with large quiffs and pretty girls with long flowing hair entering the music business without any prior experience, or the apprenticeship served by gigging around the pub and club circuits. This year, maybe as a token or as a sign of rebellion, a dog won one of the bigger talent shows! No kidding, is this the start of the animal celebrity era? I hope so as it could be extremely entertaining, and it'll certainly give a boost to Old Shep!

For my personal tastes, it occurs to me that music and performers are more talented and create more original music in America than in the UK. In the states one has to fight almost overwhelming odds to achieve a position in the music charts, and even harder to reach the potential consumers widely dispersed from huge cities to the miniscule populations inhabiting those barren and lonely areas of such a huge continent. In the UK fame can be found much easier as the population is smaller, the choice of media is restricted and it is possible for a wannabe celebrity to travel and promote themselves in virtually all corners of the country. When one achieves a Number One in the USA it is an accomplishment to be proud of. When one achieves a Number One in the UK it is most likely down to good promotion by a successful music agent or promoter or borne up to that elevated position on the back of a famed television presenter, talent show judge or personality. To back up this statement, one only has to examine the UK's music charts and note how many artists listed are actually American!

So can we define today's music industry as yet another form of mass produced, profit orientated manufacture or can we justify the direction it is taking as the correct path to suit modern tastes? If we decide that the music industry should control and guide us through our choices of musical celebrities, should we not repeat the question and enquire if there are truly any musical experts anymore? More importantly, how do we define a musical expert? I admit I have no answer to this, most of the so-called experts brave enough to appear on television over the past four decades or so have done little to restore faith in the ability of the music industry to recognise the varying tastes of the consumer. The music media has an even tighter grasp upon the throat of free music, constantly ramming what they decide is a new talent or material into our ear drums. Some even consider they know better than the record buying public. A prime example was the banning of the rock band Status Quo by the BBC's

Radio One, some over paid dimwit decided that age was far more important than the success and longitude of that esteemed band! I would ask who the hell are they to decide who I choose to listen too. I believe Cliff Richard was also banned, but I will refrain from commenting on that decision.

Can we define our music choices and musical artists down to the personal play list selection of a radio station management? Possibly but rather it seems we decide by retuning to a less opinionated radio station! Bands such as Status Quo may have received their bus passes years ago but their followers spread across generations, from the fourteen year old head banger, to the false teeth clicking pensioner rattling his walking frame in time to the beat. Why should radio stations decide I should listen to rap or R&B constantly and performed by artists that will disappear into obscurity next year? Why does much of the music media consider only the young should dictate what the audience listens to? Have they forgotten where the money is? Certainly not in the hands of school children or young teenagers, the money is in the hands of parents and adults who have gained an income of their own. Nope, I will choose what I listen to, and with little care for what the populous of trend followers consider as good music. Just call me a musical dinosaur if you wish!

Now one must ask the question, is there actually anything new in music today or has every form and style already been heard? First came Elvis, then The Beatles and in the seventies someone dredged up The Sex Pistols who were accredited with one great change in our conception of music, followed closely by the rappers who draped themselves with gold bling and used strange hand gestures to re-enforce the content of their music. Since then I do not know of any bands or musical genre that has had any significant impact on the music industry or our own individual tastes in musical entertainment. I could be wrong!

But what do we consider as a great tune or song? Do we still acclaim the songs of John Lennon and Paul McCartney of The Beatles? Perhaps those wrinkled Rolling Stones or possibly other songs regurgitated by various artists throughout the decades. Or do we consider the more recent and modern tunes to be nominated as all time greats? I daresay if one asked a group of young people and by young I mean those hidden under hoodies and under the age of twenty years old, not those middle aged musical heathens who proclaim their expertise in modern music this question, their answers would concern more recent musical performers. I may choose Elvis or Chuck Berry or possibly even Rod Stewart as the best producers of those songs that last a life time, but a younger person may think me insane. They may never even heard of these musical giants, preferring the talents of Gary Barlow, Justin Bieber (who?) or Cheryl Cole who is actually occupying the number one spot in the UK charts at this very moment of writing. Our definition of a great song or a popular artist differs amongst us all. We all have the mental ability to decide what music we prefer, those who do not have the mental ability simply follow a trend.

But where does classical music, folk music or country and western music fit into all this? I have heard on many occasions a person state that classical music is rubbish. My opinion of those persons cannot be printed here, for the sake of decency. Rubbish they call a classical symphony that has survived for well over a hundred years, rubbish they shout at Beethoven, Mozart and Schubert from the ninetieth century, or Elgar and Strauss from the early twentieth century! Rubbish they shout without even considering the hard work, long hours of practice and skills of those musicians who reproduce such complicated pieces of work. No electronic backing tracks or pre-recorded tracks litter these musical productions, this 'rubbish' is simply down to the talents of an original composer and the orchestra. But these and other such facts fall upon deaf ears (*pun!*) when many take the smallest amount of time to consider other forms of music. It is only the music or artist of the moment which captures some, not the long process of musical development over the centuries.

'So who won last years certain talent show?' I ask.

'Dunno, can't remember.' is the usual reply!

How can one call a piece of music that has endured the test of time rubbish? The very fact that it has survived the decades or centuries gives concrete proof that it is certainly not rubbish, in fact far from it. But to those who assume their own individual choice of music is the only form acceptable, in their ignorance they fail to understand the complexity, the mental capacity and the detail involved in the composition of a classical symphony or a million selling chart topper. In truth a music lover should be able to identify the structure and identity and possibly even the message in all music, the fact that a particular piece does not appeal to them as an individual should in no way deter from the fact that it is a work of art in itself.

The same applies to all styles and forms of music, from the tribal drum beats to the violin concerto; from The Bee Gees to Snoop Dog, each and every piece of music holds its own merits, its own identity and has its own followers. Even the *one hit wonders* hold their place in musical history, though more as a by-line than a paragraph. So can we define music? Certainly we can define the word music. But can we demarcate music in all its forms, styles, fashions and genres? Possibly if given enough time and thought but there is no quick or easy answer, otherwise I would have included it!

Can we delineate today's music industry? Being gracious and polite I will admit that not all music producers or companies attempt to force their chosen star or starlet down our throats, or to be more precise, in our ears but unfortunately quite a number do. I have just this moment scanned the UK's top 30 singles chart and found only seven artists or bands that I am familiar with. Who the rest are I have no idea, what their music is I have no idea. Reading some of the names used I am not even certain they are from this planet! However, in the top 30 chart they are so kudos to each and every one of them. If I could sing in any other style than that of a frog, I would hope to be in their place. If I could write a chart topping song, I would be in their place, and for the totally impossible, if I were young and very good looking I would be in their place! So I do not condemn those artists their success, but I do question their ability to survive the decades in the music industry, due mostly to the very nature of the music industry itself.

Should we then condemn the music industry? No I do not think so, the active word here is industry. Music may be the food of ones soul, but industry can put food in our bellies. Industry is described as; any economic activity that produces goods or services, or the commercial production and sale of products or merchandise or a category of business transaction. So whether we prefer the musical talents of a decomposed composer or a bling covered rapper, the music industry will produce it and we as the consumers have the choice to purchase whatever new song is flavour of the month, or run screaming away with our hands over our ears!

Like any industry even music has to turn a profit in order to survive. So if a simple but catchy tune becomes a best seller than who is to care. Not all music needs to be deep, full of emotion and with lyrics that would thrill Shakespeare. As a commodity music can be fickle and what is in vogue on this day is may be considered unfashionable the next day, or perhaps the artist upsets the tabloids and fades from popularity. The factors a music agent, producer and promoter must consider may not apply to an ordinary commodities business. We will always need food, electricity, water and of course beer. These commercial products and their retailers can demand the top market price with impunity. Housing, fuel and communications can also escape the unpredictable consumer market, but music like dining out or going to the theatre can be placed firmly in the luxury or emotional bracket of consumer interest. I mentioned the term emotional in connection with music, because it is ones emotions that lead to a certain piece of music appealing to the individual at that particular moment in time. However if one is short of cash, purchasing musical items fall short of the importance placed on food, shelter, heat or more beer. The music industry needs to keep a close eye on present trends and fashions and the desires of young people much akin to the clothing fashion industry.

The music industry is just that, an industry so we should not condemn those involved for thrusting mindless and instantly forgettable tunes into our long suffering ears. Much of the time those same gratuitous tunes are those which make the fastest profit, admittedly the fortunes of said tune may not last more than a few weeks, but in that time a music company may have received enough profit to purchase a sandwich, or luxury yacht! The artist responsible for the tune will also benefit, or at least be able to come off benefit, unemployment benefit that is. A simple song with few intelligent lyrics and a repeated electronic backing track is most likely the bread and butter of the industry. Established artists with a guaranteed sales history tend to be the icing on the huge, delicious cake for both the music company involved and of course the artist.

It must be remembered that the music industry also gave us the all time greats in musical composition, experience and performance. As an industry it has funded, promoted, prodded and cajoled new technology into existence over the years since Marconi fiddled with wires and electricity and succeeded in transmitting the first radio signal in Italy in 1895 and across the English Channel in 1899. Sound wave technology now allows doctors to treat patients without resorting to the `slice & dice' method, otherwise known as surgery, by using ultrasound methods. Hearing implants now allow the deaf to hear and as we all know, sound is used in a wide assortment of safety devices, from screaming fire alarms to honking vehicle horns, medical monitors that bleep warnings and even helps the happy anorak wandering the country side armed with a metal detector. Many of these inventions may have stemmed from the advancements in audio technology initiated either by the music industry or as a result of the knowledge gained in pumping out tunes for profit.

The music industry plays a huge part in our lives, either through entertainment, research and development or audio torture in their occasional dodgy choice of singing celebrity. In the vast majority of the music business, the consumer is allowed to make their own choice in music, style and genre, and thus it is the consumer who finally dictates the trends and fashions that others may follow. The music industry is a business, it produces a product and as such will use all its skills to ensure the business survives, and to survive it must keep us happily buying its product, music.

It is the music media I hold to fault for the majority of mindless trash we suffer as it emits from our radios and television in waves of tuneless and repetitive boredom. By music media I of course refer to the radio presenter who believes that only the young have any interest in music, or the television music presenter, or the talent show judges that attempt to promote themselves more than the wannabe artist that braves the mocking audience in their attempt to achieve a foot hold in the music business. The days of The Old Grey Whistle Test have long gone. This was a show that featured and introduced both old and new artists and allowed the public to decide. No backing track assisted boy bands or one hit wonders graced the unadorned stage of this show. Each artist had to be of such a stature in their musical abilities in order to obtain a slot on the show. Jools Holland has taken up this mantle with his Late Show, but unfortunately most of the other so called music shows only favour those instant pop stars with little actual talent but a superb promotion team behind them.

The music industry and the music media are two very different animals and should not be held jointly to blame for the weaker, less talented artists who scream from our radio's or television today. Nor can both take the lion's share of acclaim for those brilliant artists who occasionally break through the warbling and wobbling riff raff that clutters music today.

But can we define music itself? In all honesty, no we cannot. How can a selection of sounds have such a profound effect of our lives, on our souls? Some melodies can reduce one to tears, others make one smile, and of course those manufactured artists and their inane songs that make us want to regurgitate our entire meals for the day! Is this fair? No I doubt it; I have simply used examples to achieve a point. How can one define music in all its emotions, aspects, styles and genre, in all its happiness, sadness or irritation? One cannot define music in just a few words; one cannot express the emotions that music can arouse, the energy one can experience or the compulsion for movement throughout ones entire body. Nothing else produces such effects within us, nothing else affects us so deeply, and nothing else can mend a broken heart, cheer a distraught soul or bind people together like music.

We are extremely lucky to have access to such a wide variety of music and the technology to enjoy our choice of musical entertainment in our homes, workplace and our travels. Music has touched humanity like nothing else, almost every film, television programme or advert is set to a background of music. Even cows in their cowshed are alleged to enjoy music while being milked, some dogs also prefer some form of music via a radio or other such device when left alone in the home. It is considered by many to help keep the animal relaxed and calm and thus avoid the howling and barking each time the owner steps out of the house. Shame music does not have the same effect on young children!

Occasionally one will encounter a person who insists they do not like music in any shape or form, but let's be honest here, these people are either deaf and cannot hear music, or they are miserable gits who tend to complain about everything! From tribal music, to classical, from hard rock to last year's new talent, and from folk music to barbershop singers, music is now a part of our every day lives. But can one actually define music? In my opinion, no it is simply too vast a concept to define down to a single sentence.

Chapter Ten. Television

Well after rabbiting on about music and the media, I suppose I had better mention the good old television, a sort of commercial break if you like. Like others I do watch television though I try to limit the time I spend sat with my eyes glued to the black box thingy in the corner of the room. I am not a fan of most television programmes, certainly not those popular game shows, talent shows or soaps but I understand I am probably in the minority.

So what can we discuss about modern television programmes? Many people when asked state they do not watch much television as they consider it mindless and a waste of time. However when one of the less intelligent game shows, talent shows or soap opera's crop up in conversation, it soon becomes apparent that they are somewhat of an expert on these programmes. One must conclude that a small fib has been uttered and in truth these people do actually spend many hours in front of their televisions but are possibly too embarrassed to admit it. I recently asked the same question on two of the major social networking sites and received the same response. Not one comment was in favour of modern television, most stated that either the programmes did not attract them, or they felt television was simply a distraction from more important activities in life. So far no one has offered a complimentary response to my question. Why is this I wonder? Is watching television a social misdemeanour or do we secretly pretend our lives are too full and interesting to bother with the entertainment media? In truth I feel the latter is most likely the better explanation and I admit I have cast denials in the face of my actual television hours.

What ever method we use to analyse our personal intake of the magical media machine in the corner of our lounge, kitchen, bedroom or ones castle, we cannot get away from the fact that television is now an integral part of our lives. Travel on public transport if one is brave enough, and within a matter of moments one will hear a conversation concerning some form of television programme or episode, be it Eastenders, Coronation Street, Neighbours or Fireman Sam. Much of the topics discussed in the company of friends, relatives and work mates will revolve around the previous nights viewing. Matters of importance such as the economy, world peace and the price of eggs are shunned in favour of who is doing what to whom on which soap. Many people take their choice of soap opera very seriously, I once over heard a comment from one mature lady to another, I cannot remember where I heard this but it has remained in my mind ever since.

"I don't like Eastenders much. They're not real people, not like Coronation Street - at least they're real people and I like that."

Okay, so what did that lady actually mean? Did she consider that those on Eastenders were actors but those on Coronation Street were actual people going about their business? Did she assume cameras had been secreted in covert positions around each of the characters homes or place of business? Or did she simply mean she preferred Coronation Street story lines to that of Eastenders? To be honest I would not like to judge, besides the thought of Coronation Street consisting of real individuals going about their daily lives is a much more interesting concept! If cameras were hidden in homes and work places, who would edit the recordings? I know there would be plenty of volunteers for such a job, but who would ensure peeping toms, blackmailers and tax inspectors did not get the job? I also wonder but not to much, which room would receive the most views . . . ?

We all have our choices in what we decide to watch, in virtually all cases anyway. It has been known for a certain member of the family to gain possession of the television remote control and claim the right to watch programmes of their own selection and ignore the wishes of others. I have deduced this event occurs more frequently when the sport of football is featured amongst the viewing selection. It is certainly a fact that the all powerful television is held responsible for more marital and relationship destruction than almost any other factor. One may forgive ones partner for the odd affair or three, one may even forgive the spouse who inadvertently left the winning Lottery ticket in the washing machine, but changing television channels and missing that night's episode of ones favourite soap instalment is instant grounds for divorce!

One of the strangest viewing choices is that of watching programmes that portray the ordinary lives of others. Sitting glued to our armchairs and watching fading celebrities perform selections from real life that we ourselves do each day. What is the huge draw in watching a fictional family, why don't we just watch our own? Certainly the families enacted on television do tend to lead more interesting lives than the majority of we ordinary folk. Extra marital relationships; bribery, corruption, in fighting, long lost relatives showing up at the most inopportune moments, frequent fatal accidents, more secret affairs, explosions and train wrecks are all witnessed in the television soaps, but not one trip to Tesco's! How much different to our own lives are those portrayed in the programme? Actually quite a lot I would say. I have never witnessed a train crash on my door step, nor heard of anyone driving their family into a canal. I suspect the odd affair does occur in my neighbourhood but I am not saying! I do not recall ever having an unknown relative knock on my door, nor an explosion in the local pub. Why then are the fictional lives of those characters in television soaps so much more interesting than our own? Of course the answer is simple and the key word is fiction. Despite our refusing to acknowledge the fact that most of us spend the majority of evenings at home watching television, we all like to be entertained and watching our own family is hardly compelling at best. So we watch fictional characters portray fictional lives in a fictional storyline that seldom comes close to resembling our own mundane lives.

It has been observed that people in general become quickly interested in continuing story lines on a television show, no matter if the programme is a soap opera, a crime series or medical series. We soon become attached to the characters and feel a compulsion to watch and discover how each storyline pans out. The same can be said about reading a novel, one eagerly turns each page in anticipation of the next event or adventure the hero finds himself in. We imagine ourselves in his or her place, fighting off the villains, solving crimes, having that delicious affair or travelling far away to strange and exotic locations. Even the non fictional programmes are written and edited to capture our attention in this way, what is the recipe for that pudding? Can the lion cub survive those prowling hyenas? Is the house renovation going to be completed in time and will it resemble the result of a dozen drunken art students let loose with unbounded funds and extreme imaginations? I reluctantly must add that even political programmes are designed to enthral the watcher though I have no idea how this could possibly be achieved! And of course there are the sports programmes. Most of us have at least one sport we like to follow, maybe not fanatically but one we personally find entertaining. Most men it seems if one is to believe the media, favour football or rugby or cricket. Others with perhaps more freedom of will or possession of the television remote may follow motor racing, ice hockey, gurning or snooker. The selection of sports shown on many television stations attempts to comprise most viewers' tastes and just to be sure, will include the Olympics, the football world cup, and the Grand Slam tennis championships along with the golf and cricket major events. Personally I object to the absence of the tiddly-winks world championship and the unicycle team events! But what ever our choice of programme, be it soaps, sports, politics or romance or action films, each and every one is designed to hold our attention, sadly few do.

Remember my comment concerning the windscreen of a vehicle resembling a wide screen television? Well here I go again. Most of us watch too much television these days though we may deny it fervently. Could this fact be partly responsible for our impression of security and isolation when viewing the world from behind a glass screen? Nowadays as the size of television screens increase daily, is there a relationship between our vehicles windscreen and the size of television we own? I have been in several homes recently where the television is far too big for the room it is situated in. Of late I was invited into a small cottage living room with the largest flat screen television squeezed in one half of the room. I quickly got the impression of the residing family with their noses almost pressed up against the screen as they sat upon a small sofa riveted to a programme or film emanating from a marvel of modern technology only inches from their eyes. What could possibly be the attraction of owning such a huge device in such a small home? I soon discovered that the man of the house had purchased the television so all became clear, penis envy! A big motor car, a huge television and a very small home! Figure the correlation out for yourselves.

In a few homes the television is still turned off when guests arrive and is kept small and unobtrusive in a far corner of the room and often partially disguised by the careful positioning of ornaments, photo frames and vases of flowers. Sadly this trend is decaying amongst the media orientated family. When guests arrive it is not uncommon for them to be shown a seat in silence while the householder continues to watch his or her favourite programme. Only when the programme has concluded is the visitor welcomed and refreshments offered.

What is the attraction of television in modern society? Do we actually learn from it or better ourselves by prolonged viewing? It depends of course on what form of television programme we enjoy. Television may aid us in our understanding of the world, our environment and social interactions more than we realise, but of course many programmes can have the complete opposite effect and contribute to 'dumbing down' the constant viewer. Watching documentaries and factual programmes may increase our understanding of the world in general. Following quiz shows and information programmes may help in our continuing education. Certainly we get a huge amount of our information concerning world affairs, weather conditions, our economy and financial situation from watching news broadcasts on television. Without our electronic visual media we could be reduced to following rumours and gossip being passed from cities to towns onto villages and hamlets via word of mouth. But I jest as this is not a realistic prognosis; one is forgetting all the other forms of news media including that of the tabloids and broadsheet newspapers. Personal information may abound in public, if newspapers were the sole form of national and world wide news, we may even see the contents of our mobile phone texts in newsprint before we even read them on our own phones!

The Internet and all those technical devices that many idolise also supply huge amounts of news and information from every corner of the planet. Unfortunately this form of media information relies on each person actively seeking out the required information or up to date news. Yeah right! Who can imagine the average teenager or those whose lives are far too busy and important to bother with what may be happening in other countries, or what mischief the politicians have gotten up to. The main factor that contributes towards so many viewers watching news bulletins is that we are mostly too damn lazy to switch channels or face the ultimate challenge and turn the television off!

But what of game shows or soaps or action films, do they contribute to our personal development in real life? Sadly it must be admitted that they most likely do contribute in some strange way. Watching any form of television programme is likely to expand ones education and understanding. We are now familiar with the lives, accents and to a certain degree, the geography of Manchester through Coronation Street, the east end of London via Eastenders, the lives and times of Australians thanks to Neighbours and the young people of Cheshire are as shown in Holly Oaks. We learn about the national crime scene in the UK, America and Australia through the numerous 'Cops' shows and by now many of us are connoisseurs of antiques thanks to the antique and collectable related shows cluttering the television channels. Wild life abounds in our gardens and countryside and so do the wildlife shows. We are now learning how to cook exotic meals, decorate our homes, repair our homes, identify a hundred different species of birds, insects and assorted thingies that share our occupation of this planet and successfully tend our gardens with the aid of such informative programmes. OK, I say informative in the loosest possibly use of the term, but even so, these programmes must eventually help someone, somewhere to cook the perfect meal in their perfect house while surrounded by their perfect garden which contains the perfect variety of wild life.

Our understanding and awareness of medical conditions and treatment including simple First Aid and what we should do in an emergency has grown over the years, mainly thanks to such programmes as ER, St.Elsewhere, Casualty, Holby City and House. Regrettably in real life, few emergency departments are as caring, understanding and efficient as those portrayed in fictional stories. Conversely modern medical shows do pass on more information to the viewer than the medical based programmes of yesteryear or in some cases even ones actual doctor. One did not expect to learn a great deal from Dr. Finlay's Casebook or Dr Kildare in the 1960's or even Emergency Ward 10, considered by some to be television's first soap opera. Recently a programme showing the live dissection of a human cadaver was televised and viewed by millions and resulted in many diets initiated on that day! Plus an almost overloaded sewage system as hordes of viewers with delicate stomachs headed frantically for the toilet!

Children's programmes are written to educate alongside the entertainment value, so many general viewing programmes and shows drip feed the average adult little snippets of information, facts and figures that may otherwise escape their attention or understanding. I am not talking about documentaries or factual programmes here, but the light entertainment shows that clutter our TV channels. No matter how small or insignificant, a fact muttered on a popular television show is picked up by thousands without the viewer even being aware that they had just learnt something. It is a shame the same cannot always be said for the televised adverts that pollute our eyesight and intelligence constantly.

In an advert for Ski yogurts a young, bronzed and six packed male was shown to be picking strawberries off a tree! I could be wrong here but I always understood strawberries grew as plants, not trees! Sadly this type of commercial gaff is not uncommon as advertising whizz kids strive to produce the perfect marketing promotion, but lack the intelligence to check their facts first! But if one can advertise the illusion that a certain caffeine filled beverage can give one wings, or a particular male antiperspirant spray will make one irresistible to women, a simple horticultural mistake can be forgiven.

Being educated by television is one thing, but how does the magic box influence us in other ways? Do we begin to emulate our TV hero's and favourite characters? Yes many of us do, though some much more than others. Age plays a part in just how much we are influenced as individuals, the older we are the less likely we are to admire or even care what roles the actors portray on the silver screen, that is if we can see the damn thing in the first place, hearing it comes in a close second!

It is considered in many expert circles that television does affect the way in which we act or lead out our lives. The constant barrage of violence in films, on TV and now in computer games has diminished our conception of aggression, and we are seldom shocked or disgusted at scenes of death and destruction on news bulletins anymore. Young children believe that death is not permanent and we will bounce back to normal within moments, just like a cartoon character. Neither Tom nor Jerry ever actually dies, despite the bashings they rain on each other in every ten minute cartoon screening. Our acceptance of violence is evident in the gangs of thugs roaming our cities and attacking the unwary bystander or members of a rival gang, they care not which. The days of one against one fighting have long gone, now it takes a gang of more than three to savagely attack one single victim. No more is the knocked down victim allowed to regain his or her feet before the attack continues. Today once on the ground the assault on the victim continues via kicking and stomping by the gang members or similar self proclaimed *heroes*. When the victim is finally unconscious and lying torn and bleeding on the ground, the attackers girlfriends move in to rob the poor victim! Where is the pride or self esteem in such a one sided fight? There is none, none at all! So why are such attacks becoming more common place? Possibly because too many do-gooder people who live in large houses away from council or housing estates still insist the offender is more important than the victim. But lets not dwell on this gruesome fact, the subject of crime and non-punishment is one I do not wish to pursue. For now let us continue with the subject of our beloved television and its affects upon our dreary lives.

Television and film personalities have the greatest impact on many of us as we unwittingly attempt to model ourselves on our screen hero or heroine. Young women starve themselves in an effort to slim down and reach that exalted size zero but instead they find themselves in the grip of bulimia. This eating disorder involves binge eating followed immediately by purging the contents of ones stomach before any calories can hit the waistline. Or the other main concern of too much weight loss, Anorexia where the suffer restricts their food intake to dangerous levels in the fear of becoming overweight, even though they are usually already seriously under weight. Recent research has discovered eating disorders are increasing amongst young men and concerns have been raised.

It is well recorded that super skinny models and film stars contribute to both these illnesses especially amongst young females. In men it is often the powerful urge to become a rippling, bronzed muscle bound Adonis like the male models strutting across their television screen like pampered peacocks! Neither of these body forms is true to life, only to that of dreams. A super skinny woman is simply not attractive to a red blooded male who wants curves and a figure he can hold on to, not a stick insect that would slip from his grasp like a eel from a fisherman's hands. As a male I enjoy seeing curves on a woman, not sticking out ribs or legs so thin and bandy one could drive a bus between them! Many women (*so I am told*) do not find an overdose of muscle attractive either, I certainly do not fit this criteria! So why do people go to extremes to achieve such a body? Is it because their favourite personalities of the media circus lead them to do so by example? Unfortunately this is often the case.

The natural shape of our bodies is designed by nature and as a result of our individual lifestyle, not by a skinny celebrity or a bulky muscled film star. OK, let's be honest here, muffin tops, beer guts, double chins and love handles have adorned our bodies for centuries. Today machines remove the hard labour from our employment, food becomes more and more available and our entertainment revolves more and more around technological devices, our bodies have little exercise to combat the production of weight gain. As a race we are becoming fatter! I certainly fit well into this group.

Viewers can also be influenced by consumer items rather than personalities. I refer now to the over indulgent food adverts crammed onto our screens at every possible opportunity. With such tempting delights ready made for consumption constantly on our television sets, it must be considered prudent that *smell-a-vision* is not yet available or we would have no chance! Not only seeing mass adverts for tasty morsels, Sunday roasts or chocolates by the dozen, what if we could smell the food being advertised as well? Disaster! There must be a limit; curves should not be allowed to develop into small hillocks, deep folds of fat and flapping tissue carrying the danger of small children and animals being lost forever in the obesity of human excess.

These two examples can be pictured sharing the delights of an eating establishment. On one side of the restaurant we have the skinny wannabe models nibbling on a single leaf of lettuce. On the other side of the restaurant we have the hamburger stuffing, chocolate gulping, French fries devouring corpulent individual. Both in their own way have been influenced but what they see and hear on the tantalising but intimidating television screen. One has been coerced into desiring the perfect body, the other has been seduced into self indulgent extremes by the wanton display of edible luxuries displayed in profusion in the advertising media.

Television advertising is considered partly at fault for those morbidly obese individuals, though it does not hold the blame alone. Television is not responsible for MacDonald's or Burger King or Pizza Hut or Kentucky Fried Chicken or even the lowly fish and chips, but it does constantly remind us of their existence. Supermarkets and shopping trends also lead us all towards the convenience foods, the high calorie foods rich in sugars or fats. Yummy! Yep we all love those foods and our television not only reminds each one of us on a daily basis of the delights on offer, but also instructs us on where and how to obtain these saliva inducing products. The food industry contributes millions of pounds and dollars to media advertising so few can complain, after all no one forces us to stuff goodies down our throats at every opportunity or to starve ourselves to almost the point of invisibility.

In truth however, the benign flickering silver screen positioned prominently within our abodes does have a huge effect on how we see ourselves, be it in a double wide mirror or the thin reflective handle of a tea spoon. Our very own individual characters may become changed by those we watch and admire on the big screen or the not quite so big screen, film and TV to be precise. Like our growing acceptance of violence for example; or the desire to emulate the figure of a television idol, possibly even a fictional film hero. Most of us fail to recognise this change, perhaps it occurs subconsciously at first before spreading throughout our whole existence. Our media heroes can attach themselves to our persona as psychologically our characteristics mirror those we admire, and hence we become more like our favourite media characters than our own actual personality.

In order for an audience to be successfully influenced by a television programme or big screen film, the media itself must be successful in its ambition. So what does make a good programme, piece of entertainment or movie production? This is a very difficult question and if the answer became apparent, every media company in the world would flock to obtain it. Our tastes in television programmes are as varied as our tastes in life. This is what sets the human race up as the dominant shopper in nature; otherwise could we compete with the animal kingdom in making sensible selections of purchases? Probably not as most animals already have a better dress sense. A zebra looks fine in stripes and a leopard clearly shows off its spots. But a human dressed in one of those virtually fluorescent track or shell or sport suits could shame even a hyena! Animals also have a better understanding of diets, a lion remains on a protein diet similar to the Atkins diet of limited carbs and high protein, while a gazelle remains on a strict vegetarian diet, no huge hamburgers for them!

But being serious, to gain our interest and thus achieve some form of influence upon our choice of viewing, the media industry must ensure its programme content is suited to our modern tastes and preferences. This is serious, how can one truly justify the present programme selection? Do the media see us as celebrity loving, decorating, wannabe chefs who spend every free moment in the garden or chasing wild life? Strange but it must be accepted that those working in the media industry are the same as the great viewing public. They too head for their homes at the end of the day and often fester in front of their television sets. So maybe it is not our choice of programme they seek, but instead choose what they enjoy watching after a hard day in the studio? However it is possible they are simply human, just like the rest of us animals and enjoy or dislike the same variety of fiction or fact or sport that we do. The only different factor between the media producer and the common man is profit!

The drive behind our television industry including all the films, documentaries, reality shows, talentless shows and even advertising is that they need to gain a profit. We as the viewer simply seek entertainment of our own individual choice, but the people responsible for what we watch have to include an income from their programmes. Does this mean that a typical BBC period drama will out sell one of the popular quiz shows? Perhaps it will. But will the period drama out sell a football match? No, probably not. Will a talent show aimed at the younger person attract more viewers than almost any other television programme? Unfortunately the answer is frequently yes but excluding our traditional favourites, the soap operas. So who chooses what we watch? Actually in a strange and perhaps slightly unsettling way, we ourselves chose the programme list or the film subject. Does that say a lot about us as intelligent human beings? Again sadly the answer is yes. As with all industries, the media industry is there to make a profit, to make money and grow. It would not be good business to pour money into a project that was doomed to fail from the start, only governments and councils do that sort of thing! No the media industry gives us what we want, each and every one of us that is. It cannot be helped if one hates a particular form of television show, others may enjoy it immensely. Some may dislike football but many more are avid followers of the sport. Erstwhile viewers may find soap operas hard to stomach but actual viewing numbers tell a different story.

So it would appear that although we are indeed influenced to a high degree by television, we ourselves manipulate what programmes the television company's produce and what style of show we like to watch. Sadly it is our fault that ones evening viewing is crammed with celebrity game or quiz shows, our fault that that our music industry is being overwhelmed with talent show winners, our fault that every football match in creation is screened into our homes, and it is our fault that we sit and watch programmes portraying the fictional lives of pubs, streets or other such venues in soap operas. I have no doubt that sometime in the future our tastes in television programmes will change, in favour of what I do not know. But what I do know is that the programme producers and planners will change along with us. Sadly there is no escape!

Television today has become an addiction to many, the desire to watch ones favourite programme and the compulsion to interact or emulate our favourite celebrity has gained the addition to our growing list of personal vices.

Chapter Eleven. Vices we love and hate.

Think of all those things in life you enjoy, not those idealistic visions of mist covered mountains or lush green valleys glittering in the warm summer sunshine. A calm sea with empty beer cans and condoms floating on the surface, tenderly moved by a gentle breeze amid screams of seagulls soaring through the cloudless skies. No not those impressions of peace and serenity, but the pleasures we enjoy as individuals, those sources of enjoyment or satisfaction that may appear anti-social or be considered as a bad habit and but also conceivably a fashionable trend. Those small pass times or distractions that many of us enjoy in public or in private, or both if one is not bashful or concerned about the opinions of others. Those opinionated others who hide their individual pleasures under a cloak of secrecy, denying their own character flaws in order to appear superior in society, perchance only showing themselves as supercilious instead.

Neither am I referring to obsessions, though it is plausible that this term applies more than others. Nor am I intending to comment on excessive addiction, this is not a subject for a layman to contemplate here. This chapter concerns the addiction to certain vices, not full on addiction illnesses and the poor people who suffer from them. People such as alcoholics, serious drug addicts and kleptomaniacs. Gambling addiction can be considered amongst these more acute concerns, though who ever decided to name such an addiction with the ludicrous title of Ludomania I will never understand! Or those people who experience the similar conditions of craving, compulsion and repetition. These are serious addictions and will always require our understanding and sympathy and cannot and should not be discussed lightly. Yet, it must be appreciated that the vices to be discussed here are now, or will become addictive for many less fortunate participants. But it is our enjoyment of certain vices that concerns this chapter, vices such a binge drinking, smoking, possibly even some of the seven major sins like gluttony and vanity. It is also how others view ones personal choice of vice, possibly with distain, with horror, with shame or with pride.

Examining personal vices, those often enjoyable practices we develop that are deemed by some as unhealthy, antisocial and more of a nasty habit than a mental or psychological problem. The more common vices are recognised by all, smoking, drinking, over eating, extreme computer gaming, fashion and of course, greed.

To begin our little investigation of human vices, we start with the consumption of alcohol for pleasure, though why such a statement exists has not yet been explained, why would one consume alcohol for displeasure? The real question is why has the UK become the centre for weekend or holiday binge drinking, drunkenness and the associated embarrassing behaviour? Has drinking become a twenty-first century fashion that all too many follow? Almost every television news bulletin shows graphic images of young and the not so young people falling about our streets following ejection in the early hours of a morning from night clubs or public houses. Large groups of revellers aged from teenager to twenties and thirties gather on beaches and secluded venues across the country to drink themselves senseless.

Those without access to beauty locations to despoil seek out street corners, car parks or shopping centres where they can partake of their vice without consideration for others or fear of being interrupted by the ever diminishing police force.

It is not just the younger generation that enjoy the vice of drinking, older and supposedly wiser drinkers can be found swaggering and swaying from one public house to another, vocalising their alcohol induced happiness in tuneless and loud song. Even the alleged social elite appear to be fully signed up members of the drinking classes, spewing out from hotels and wine bars and attempting to demonstrate their own importance by howling loudly for a taxi or personal chauffeur. Work parties, stag nights, hen nights or same sex nights all appear to follow the unwritten law of compulsory alcohol abuse and noise pollution. In truth one may forgive such a crowd of drunken miscreants as they at least have a reason for their jollities, sadly the majority of inebriated gatherings are simply for the sole reason of consuming as much alcohol as humanly possible or not possible as is often the case!

Even in the privacy of ones home the increase in alcohol consumption is apparent. Cheap booze from the corner shop or local supermarket can be carted home by the crateful for the evening's debauchery, almost always accompanied by loud music, loud voices, vomiting and irate neighbours. Especially those not invited to participate!

So why has this culture of excessive drunkenness become fashionable? Can these people honestly state they are enjoying themselves? Have we become such a miserable bunch of Brits that we require vast quantities of fermented hops or grapes in order to have fun? Do the inhabitants of other countries suffer this drain on the family or individual financial income as we do? And can the popularity of binge drinking be identified as peer related or a social concern or just another vice that many conform to?

It is highly possible that the low cost of alcohol these days contributes to over consumption, though I do not think this really applies to the social scene as nite clubs charge what ever they consider acceptable to their frequently already drunk customers. Supermarkets and grocery shops have assisted the growth of street corner drinking as week after week alcohol prices are slashed in special deals, buy one get one free or half price sales. It is strange, and somewhat disconcerting that alcohol is commonly more affordable and more accessible than food in some areas of our inebriated country.

The desire to attract a sexual partner must be considered when examining social entertainment in any country. Unfortunately in the UK, the search for that perfect partner or one night stand is constantly viewed through beer goggles, leading to guilt, disappointment or pure horror in the revealing light of day. Regrettably those men seeking female attraction with the aid of alcohol may only achieve disappointment as most beers *do* reach those parts essential for sexual conquest!

The near total lack of female clothing accompanies the rapid consumption of alcohol, girls seeking to show as much of their bodies as decently will allow. I will admit these skimpily dressed young women can be very appealing, until they begin the alcohol ritual of fighting, uttering offensive language at an extremely high decibel level or falling off their high heels, urinating in the street and regurgitating vast quantities of that night's alcohol consumption. Even the prettiest young lady will lose some of her charm with vomit dripping down the front of her expensive designer dress!

These young ladies are constantly surrounded by hordes of men all dressed in the uniform of drunkenness, the unimaginative shirt, jeans and trainers or sneakers. One looks backward in time to the high and sometimes outrageous fashions of the sixties, colour and patterns abounded, now we get shirts and jeans! Something wrong here!

The men are no different to the ladies, at first appearing or attempting to appear suave and sophisticated, intelligent and the perfect example of the male species. This façade does not remain dominant for long; all too soon the gel haired, deodorant drenched would be dominant male degenerates into a shouting, aggressive and swaggering lout with false ideas of grandeur. That figure who represented the ideal partner and potential father an hour ago, that alleged female magnet gradually disappears and is replaced by a loose shirt tailed Neanderthal with dark stains growing larger on the upper parts of his jeans!

This new race of sub human's crowd into the streets of towns and cities in the early hours of the morning as the night clubs close for recuperation. Hostilities break out as one young hero decides another has paid too much attention to his intended partner for the night. Busting bladders create streams down pavements and fountain from shop door ways and alleyways. Taxi's and private hire cars cause chaos in the pre-dawn hours as they battle to collect their designated customer and discourage the opportunist in search of a free ride home. Those lucky enough to have achieved their ambition of acquiring a sexual partner head off to their homes, vehicles or discreet hedgerows to complete their night's activities.

 Finally with the aid of too few police, ambulances and a multitude of taxi's the crowd thins and eventually disperses to other locations, homes or and assorted locations to sleep off the vast consumption of the night. All that remains as evidence of the night's licentiousness is the huge amounts of litter, empty burger and curry trays, beer cans, vomit, urine and sometimes even faeces covering the streets and pavements like the aftermath of a ticker tape parade in a landfill site.

 It does appear to be fashionable to attend such gatherings at weekends and certain holidays throughout the year. One only has to listen to the boasts and imaginative reports of the evening delivered the following day by those sober enough to still believe they only drink in order to enjoy themselves. The answer to one question is constantly quoted by a majority of established drinkers. They feel they cannot let themselves go and really enjoy themselves without the aid of alcohol. It is a necessity in the pursuit of enjoyment and pleasure that these people require the boost to their egos and to dampen their lack of confidence, build their self esteem and make to them instantly attractive to the opposite sex. Is this really the case? Is alcohol a wonder drug that can achieve all these criteria? No of course not. Alcohol could be defined as the illusionist drug, it fools one into believing what one desires and allows those partaking to escape the normality of their individual lives.

But one of the over ridding factors responsible for excessive alcohol consumption is peer pressure. If ones friends constantly stagger about under the influence and swear each had the time of their lives and continually brag about their personal capacities for consumption, then one is compelled to follow their example. As more and more of those under the age of forty seek to emulate this lifestyle, so the younger generations will follow. Eventually we see children not yet in their teens lying in drunken stupors in parks and on beaches across the country as they too endeavour to follow their elders and peers. Sadly this event is now already a regular occurrence.

So it can be concluded that in some abstract way, the high consumption of alcohol is decidedly fashionable at this moment in time, but one wonders if the end result of such excessive consumption really creates pleasurable memories. Having personally experienced the odd such night, weekend and occasional week, perhaps the end does justify the means. Alcohol is a vice that the majority of us enjoy at some time or another. In the past, America attempted to ban the consumption of alcohol and soon discovered their folly. Alcohol is a socially accepted vice shared by the majority, it can be a release from the stresses of life, an antidote to low self esteem or an enhancement of confidence. It can provide humour in action, thought or consequence, but it can bring forth sorrow despair and regret. It can turn a super wimp into a superman within hours and it provides the perfect excuse for any outrageous or excessive behaviour. Drunks fall heavily on the excuse of amnesia to avoid explaining their antics of the previous night.

The majority of people staggering about the western hemisphere partake of the odd drink with impunity from the stigma attached to smokers. No stigma perhaps but the health risks are just as high and do not solely apply to the heavy drinkers, binge drinkers or alcoholics. The risks include all forms of drinker, be it the glass of wine with a meal, a snort of whisky in the evening, a tot of brandy beside a roaring fire or a gallon of cider on a hot sunny afternoon. And as for home brew, well the least said the better! Alcohol related diseases include; cancer, impotence - now there's a surprise! Others include diabetes, high blood pressure, cardiac disease, cirrhosis, anaemia, pancreatitis and diarrhoea.

While the consumption of alcohol is more acceptable and fashionable than the social drugs, it is still a major risk to health and is the root of many petty crimes and violence. Alcohol is regrettably a vice that some over partake too often and make choices in their lives that would certainly not have been considered while sober. Alcohol is an accepted habit for the majority of consumers but it must be remembered that alcohol is also classed as one of the social drugs!

The popularity of drugs has increased over the decades, despite all the warnings and available information against their use. Social drugs such as heroin, cocaine, barbiturates, street methadone, benzodiazepines and amphetamines are all readily available today. It must also be recognised that many of the mentioned drugs are used by a wide cross section of the population, not just addicts shooting up in dark alleyways. Cocaine is widely used in the more affluent circles and several named celebrities have fallen foul of its use. Across younger age groups, drugs are used to heighten ones enjoyment of a night out in a dance or night club and especially in the illegal but popular raves, a gathering of persons in a secretive venue for the purposes of drinking and dancing with the odd fornication amongst the bushes.

Although in the main, drugs of any kind are frowned upon as anti-social, unlawful and a serious risk to health, many still pursue this form of vice or habit within their social circles. It is a fact that while a drug user may shun attention and not want to be seen rolling a joint, snorting a line or shooting up in public, the practice is still commonly accepted in certain social societies. With the growing awareness and understanding of the damage these so called social drugs can inflict, the main perception of public opinion of these drugs has obtained the distinction of being socially unacceptable, however it certain situations they are still seen as fashionable, unfortunately!

Tobacco! Now here's a taboo subject and vice to dwell on for a moment or three. I have no doubt many reading this will have strong opinions on the evil weed, but is it justified? OK, slow down a moment, yes we have all heard the dire warnings from the medical establishment, but as one of our every day vices, tobacco and the choice whether to smoke or not smoke still constitutes a human dilemma so we will include it in this section. Unlike many of our other long list of vices, tobacco and its use has encroached into the realms of fashion rather than health. It is now socially unacceptable to inhale a noxious concoction of smoke and chemicals into our lungs and our immediate environment. It all sounds nasty but are our concerns truly centred on health, or a modern trend? Of course many will consider the health complications, but do we all? The threat of passive smoking resides high in the minds of those forced to breathe the same air as a smoker. But what about the drunk driver, is he or she not a serious hazard to the health of themselves and others? In fact many of our other vices are also a threat to our health but we choose to ignore these. Has the decline in smoking really come around as a result of personal health or is it simply not fashionable any more? Personal health or fashion trend?

In the early movies of the twentieth century, cigarettes and tobacco were considered as a fashion accessory, all stars of the silver screen could be viewed happily chuffing away on a nicotine stick. It was acceptable to smoke in all public areas and those who did not smoke were invariably viewed with suspicion. One of the most famous figures associated with smoking in the UK was Winston Churchill. His famous cigar held in plain view between two fingers became a symbol of victory during the dark years of World War Two.

Tobacco is believed to have been in use since around 3000 BC but as smoking is recognised as being bad for ones health, there is no one from that period alive today to vouch for this, so this fact may be false. It was introduced into Eurasia in the sixteenth century and became a common commodity on the major trade routes of that period. However even at that early stage of tobacco's attempts at world domination, some brave hearts were initiating the fight against it. James 1st was particularly against the use of tobacco and is noted as stating tobacco as harmful to the eyes, brain, lungs and nose and compared the smell to that of a hell pit.

The link between smoking and cancer was first identified in the 1920's by German scientists but this fact was overwhelmed by other factors dominating that period of history. The British finally figured it out in the 1950's but took another thirty years before declaring a distinct link between smoking and cancer. The abundance of information linking smoking to all things evil has resulted in a decline in the practice of smoking since the 1980's.

Okay, so enough with the facts. The item on discussion here is why the vice of smoking is suddenly considered unhealthy and antisocial or could it better be described as unfashionable? It is well considered that smoking is bad for ones health but that is not what is being referred to here. Smoking actually appears to have become unfashionable rather than unhealthy. If one is solely concerned about ones health, other factors such as driving too fast, unhealthy eating and an almost total lack of exercise should also be considered unfashionable. If all those who describe smoking as a risk to their health considered other common health issues as well, like obesity or alcohol or narcotics, a different perspective on smoking may be considered. There are no favourable votes I can add for the vice of smoking as there are simply too many negative votes associated with it. However if one is interested in personal health issues, why do we not place as much importance on keeping fit, a healthy diet and low alcohol consumption?

Smokers these days are ostracised and forced kicking, screaming and coughing out of public buildings and into cold damp shed like structures outside pubs, clubs and restaurants while those who do not smoke continue to enjoy the comfort and freedom of a pleasant indoor environment. What about everyone's human right? Why should we condemn those who smoke to an inferior life style? Actually the answers to these questions are easily established. One individual smoker in a packed room can affect those around them via passive smoking, thus forcing their personal inclination onto the lives of others.

If someone wishes to over eat and carry sack loads of extra weight around with them, that is their choice and it affects no one else. Or does it? What about the cost to the health services in treating obese persons? We all have to pay taxes that contribute towards the treatment of the health issues that blight the morbidly overweight so does that not also infringe on our lives? Of course it does however it only has an affect on our pockets, not our health. Sorry but in a way, the vice of gluttony or wanton obesity can affect our well being by reducing the resources health organisations have available for the rest of us.

Although I appear to be unsympathetic towards those extra, extra large members of our society it is not my intention, it would not be fair to label all weight concerns as self inflicted. There are many medical and psychological reasons for weight increases, steroids being an example in question, specifically the anti-inflammatory drug know as Prednisolone. Diabetes can be a cause of weight increase as can glandular health conditions along with many others. Mental stress, anxiety and depression may also lead to the suffer putting on extra pounds as they battle against sometimes overwhelming odds.

Weight increase and weight loss can not be included in our modern collection of vices, and I must watch what I say here because I myself am no picture of health, vigour and six pack. Nevertheless though not strictly a vice, I do consider the growing concern of obesity has an impact of our society and in some situations, it can influence our personal lives, but only if we ourselves intentionally allow our weight to become a problem. Alas there are those people who for a variety of reasons cannot control their weight, however for many others it is simply down to eating more than they need and exercising less than they should.

The following of fashion in name brand clothing, footwear, perfume or accessories such as designer handbags could be described as an addiction amongst many in this celebrity lead world. Perhaps the vice of vanity! People spending hard earned and sometimes meagre incomes just to own and display the same footwear as their favourite television personality, the same hair style as a top footballer or simply the same brand of clothing favoured at that moment by ones peers. The latter has far reaching concerns, especially amongst the school age youngsters. Not wearing the correctly branded apparel is known to give rise to bullying in schools and occasionally leads to being ostracised within ones own community.

Fashion in all its forms appears to be on the increase as the electronic media constantly flashes images of alleged visions of beauty that we should all strive to emulate in our dire and bleak lives. It is strange that those seeking to follow a certain fashion craze that will display their individuality to the world, often fail miserably. If one person initiates an original vogue, especially if supported by a popular celebrity, this fad is immediately copied by thousands, even millions. In a very short time, this once unique trend has given rise to a sheep like similarity amongst fellow fashion followers.

Children plague parents for the latest fashion of school skirt because they desperately wish to fit in with their peers or possibly, at the beginning, wish to out shine their peers. In a period of only days, girls throughout that particular school have beaten their parents into submission and acquired the exact same item of clothing, often at some reduction to the household finances. Suddenly that one girl with ambitions to be a trend setter becomes lost in the tide of same style clothing that has now flooded the populous of girls attending that school. Like a plague this obsession with following the herd ensures that in a very short time all surrounding schools are swamped with young girls resembling clones. Originality has passed its sell by date!

Sadly for those whose parents or guardians are too poor or those unfortunates in care quickly become the subject of ridicule as no funding or opportunity allows them to purchase the new fashion must have. It is a fact of nature that adolescents are governed by the desire to follow their peers, it has long been established that the vice of smoking frequently originates within the latter years of school. Binge drinking appears during those sixth form or college semesters and progresses into early adult life. So fashion itself can be included amongst the vices we all face on a daily basis but remains one of the more acceptable addictions in the eyes and minds of the generations.

Vanity in the following of fashion at any cost is obviously not in the same league as smoking, over indulgent eating or the consumption of alcohol, but when one is the target of abuse, bullying or ridicule then fashion itself can stray into a threat.

Returning to alcohol versus tobacco, consider then the relatively new vice of binge drinking, the apparent desire to become as inebriated as quickly as possible with no concerns about health or wealth. Why is being extremely intoxicated seen as more acceptable than the individual smoker? One alcohol sodden individual can wreck a party, do untold damage to property, become a dangerous threat to those around them and have a negative influence on all those in the vicinity. Drinking lies behind a multitude of driving related deaths, violence induced deaths and of course, illness and early death to the individual themselves. So why do we revile the smoker but accept the drunk? Why is it seen as an obligation to become so incapable through the vast ingestion of alcohol when socialising? And why is this practice considered so fashionable that ones consumption of alcohol is the subject for the following days bragging rights? One seldom if ever hears a person boosting that they had managed to smoke two packets of cigarettes the previous day, but it is common to hear about the quantities of alcohol drunk and the resulting actions of the individual while under the influence of drink.

Possibly the answer to the distinction between a smoker and a drinker lies with our own perspective of each issue. Drinking is considered a fun and pleasurable pastime which most consider has no effect on others. Smoking is considered a disdainful and unclean pass time which is well understood to have detrimental effects on the health of the nicotine partaker and perhaps affect the health of others.

In truth though, both vices have dramatic effects on the health of the user and others in the vicinity.

Drinking has an affect on all of us, much more so than the solitary smoker sheltering from the rain in a quiet doorway. When has a smoker returning home late at night woken all within the neighbourhood? Possibly some may be disturbed by the coughing, hawking or choking emitting from that individual, but can this disturb our sleep as dramatically as the group of inebriated revellers returning home or swaying past our doors and windows in the wee small hours of the morning? Loud voices, snatches of acapella singing at the decibel level of a jet engine. Bins and other assorted inanimate objects being kicked around the streets, car alarms being set off, and a constant barrage of foul language that offends our tender ears. Those under the influence of alcohol appear to suffer the infliction of deafness, voices growing louder as the night progresses as if the speaker is attempting to hold back the silence of the night with a stream of incoherent conversation and ribald comments. All too often the noise nuisance increasing to ear shattering levels as an inebriated reveller manages to fall over a vicious blade of grass or sinister matchstick laying in ambush upon the ground. And the disturbance does not end there. Once inside the chosen residence, the inebriated gaggle of socially accepted drinkers surrender to the compulsion of loud music played at full volume while voices are raised even further in a vain attempt to be heard above the blaring beat. Finally in a last desperate attempt to broadcast to all in the area that the participants have had the odd drink and are thus supermen, several revellers vacate the residence with a profusion of slammed doors and shouted exchanges as they stagger towards their vehicles with the intention of driving home. It is a fact that most drinkers consider themselves to gain greater driving abilities and fail to comprehend their limitations when under the influence of a nights alcohol guzzling.

In truth, which would you prefer to have as a next door neighbour, a smoker or a drinker? I know which I would choose, I already have had experience of the latter! So while the vice of smoking grows in its social unacceptability, the vices of drugs and drink thrive in our lives. Smoking has lately received such a blow to its popularity that even the highly valuable antique can become worthless over night simply because it is considered as a smoking related item. The minor fact that the item is two hundred years old, was designed and built by a recognised craftsman and until recently was worth hundreds of pounds is ignored because fashion states that smoking is unacceptable. Cigarettes are now hidden from view in our shops while alcohol is displayed prominently upon over stocked shelves and high fat foods and ready meals are thrust to the fore to ensure we purchase our fill.

It may appear strange to hear arguments in favour of smoking tobacco but this is not the intension. The subject concerns not the physical act of smoking but the fact that fashion rather than health concerns have blighted our conception of this terrible penchant. Smoking of course is a serious and harmful vice; many illnesses and disease are now accredited to this practice. Cancer of course is the most commonly recognised, but respiratory conditions such as Emphysema and Chronic Obstructive Pulmonary Disease (COPD) are increasingly prevalent in modern society. The medical profession can and does often exaggerate the harm caused by smoking, possibly even laying the blame on the habit of smoking to cover up or pass off conditions of health that have not yet been fully explained. But the fact cannot be ignored; smoking is bad for ones health.

We all realise the health hazards we face if we choose to take up the evil weed and we all know or know of someone who has suffered the ultimate fate as a direct result of tobacco consumption. It simply cannot be denied that smoking is a life threatening vice and no one should even consider taking up this habit at all. However other vices can also be just as harmful so why do we have a greater level of acceptance for alcohol abuse, social drugs or even violence and crime?

When questioned, most smokers will admit to taking up the habit at a young age, many were influenced by relatives, friends, work mates and peers. Nearly all smokers agree they desire to cease the habit but find the addiction too high an obstacle to surmount. A vast majority of smokers today suffer acute embarrassment when discovered having a quiet drag in some discrete corner or even in their own vehicles, and most detest the habit and would seek to cease if they could. However giving up smoking is extremely difficult and only viewed as an easy task by those who have never smoked. Smoking is an addiction and all intelligent persons should now have a clear understanding of how difficult it is to overcome any form of addiction. It must also be remembered that tobacco is a drug and as a drug it presents its own catalogue of issues that have a deep impact on the user. The vice of smoking was deemed acceptable and even beneficial in past years; films stars, pop stars and smoking rabbits littered our screens and advertising billboards. Now in the relatively short space of two or three decades, smokers have become social lepers.

Weighing up all the factors involved with our routine habits and vices, it becomes clear that of all our personal traits, our list of sins and behaviour, we differentiate between them on the grounds of social acceptability rather than on health alone. Binge drinking is still high on the list of favoured past times and thus is viewed kindly by most. The resulting health issues, the untold number of related deaths, including those of innocents who just happened to be in the vicinity of the drunken driver or the drunken street brawl are filed under; *It won't happen to me!* Personal health deterioration due to alcohol is regularly ignored until too late, all too frequently resulting in the death of that unsuspecting but nicely pickled alcohol consumer.

Vanity and the almost addictive following of fashion is not on our list of priorities but it should be considered as a vice, albeit a small one which only affects those who lack the income to maintain the standard set by their peers or favourite celebrity. Social drugs certainly are a serious vice that we all must conclude is a life destroying pastime which when allowed to get out of hand will quickly lead to addiction.

Obesity is another growing concern *(pardon pun)* that is linked to our very modern way of living. The availability of fatty and sugary foods along with the fast food industry has been creeping up on us for decades. Increasing the portions served to customers as a lure to entice them into burger; chicken, curry and pizza restaurants originated in American. This was a successful marketing idea that followed the austerity of the Second World War. Low cost high fat meals are now the mainstay for a large percentage of the affluent world.

But can obesity be considered as a major vice? The conclusion here must be twofold. If the weight increase is wanton and deliberate or as a result of an unhealthy and indifferent life style, then yes, obesity should be classed as a vice. However if the weight gain has resulted through other factors, medical, mental or possibly just a poor income, then it should not be considered a vice but the result of extenuating circumstances.

The habit of smoking is not considered a favoured pass time by many, in fact even those who partake of the evil weed resent the addiction that binds them to this vice. Many tobacco users die each year due to smoking related diseases and others suffer years of illness through respiratory difficulties. But one must consider the affect on others created by smoking. Yes passive smoking can cause concern to those in the immediate vicinity of the smoker; this issue is easily remedied by the smoker not pursuing the habit in the company of others. Noticeably deaths and injuries affecting innocents rarely arise as the direct result of a smoker sucking away on a nicotine stick in the cold and draft of a doorway, alley or street corner.

Finally, it should be noted that smoking is generally on the decline while alcohol consumption, especially amongst the young is on the increase. Smoking has become antisocial and unfashionable while the ingestion of alcohol remains both acceptable and even an essential part ingredient of socialised civilisation. Of the two main vices, it can be concluded that alcohol is the greatest threat to the highest number of people and is on the increase. Smoking tobacco is a considerable threat to the individual alone but is judged to be on the decline as it sinks out of modern fashion and social acceptability.

The main two vices of modern society have covered the majority of content in the chapter, apart from a small foray into the vices of social narcotics, vanity and celebrity induced fashion due largely to the severity of the two vices themselves. The human being will always be susceptible to vices, habits, obsessions and addictions, drawn into these behavioural patterns by our very human nature. So do we simply embrace them or shun them? That is a question for each individual to decide. But until then up goes the familiar cry from throats young and old to savour the holidays, weekends, parties or gatherings when the thoughts of many become vocalised with;

'Sex, drugs and Rock n Roll!'

Chapter Twelve: And so to summarise.

A wide variety of subjects have been examined and discussed in this book, and conclusions and suppositions made along the way. As is always the case when opinions are given, some may agree while others will definitely not agree and it is understood that the final destination of this book may be a waste bin or lying forlornly outside the window through which it was thrown.

The subjects chosen in each chapter are not those requiring deep philosophical analyse, nor lengthy sociology investigation or psychological examination. The issues discussed are purely those of modern day living, things that we all do and see in our daily lives. We see changes all around us, some we acknowledge, some we fail to understand and others creep into our lives stealthily and un-noticed. Changes continue throughout our lives and affect us in many different ways. Some changes are important, having a huge impact on humanity, some like the hot chocolate drink is insignificant and only noticed by the individuals themselves. As a matter of interest, the hot chocolate drink has now been replaced by a milk and malt based beverage known as Horlicks. Strange I know; this vast choice of nightly beverages one consumes as one gets older. What is next one wonders, tea possibly but certainly not coffee, plain hot milk or something stronger, we will just have to wait and see.

And what about the poor defenceless Brussels sprout? This poor vegetable has long suffered the indignity of being pushed to one side upon the dinner plate, until our taste buds change with age and it finally becomes acceptable. This simple change may be of no significance in the wider scheme of the universe and the meaning of life (*42*) but it does indicate in its own small way the changes we all see in our own existence.

Changes in technology over the decades have brought us the motor vehicle to transport us in comfort, reasonable safety and financial ruin to whatever destination we desire. Form a work tool to a status symbol, from convenience to necessity, the motor vehicle rules our lives. We can now take a train or drive, float or fly to any part of the world we desire in relative speed and security which is very nice of course. Black ribbons encircle the globe and allow us the space to expand, unless one meets one of the millions of drivers who inspire road rage on a colossal scale. Or we have the misfortune to begin our trip in time for the dreaded school run, or even find ourselves sandwiched between two white vans at the traffic lights. The motor vehicle in all its shapes and sizes, function or folly has become a mainstay of our modern way of life, and most of us could not manage without it. The motor vehicle, automobile or car is especially essential for one of our new pastimes, out of town shopping!

Shopping amidst the tidal wave of humanity has reached incredible heights of diversity and success as retail outlets selling all possible items for purchase march across our land. Shopping is now as popular as or perhaps even more so than the traditional family day out. No longer do many of us search out hidden beauty spots in the countryside, or dive head first into the waters of a picturesque beach or indulge in long walks along perilous cliff tops with the wind trying hard to ruffle our lacquered or gelled hair. Once the family car is jam packed with grannies, kids, dogs, mothers, fathers and occasionally the odd innocent passerby, off we head to the supermarket or shopping centre of our choice. The more adventurous amongst us may head for garden centres or DIY stores, possible with a brief tour of a furniture or electrical store upon route. Shopping is now seen as a main form of entertainment that can rival the attractions offered by theme parks, zoos and adventure playgrounds. Shopping is *(usually)* a non-threatening experience which we all must undertake at some point in our lives. And for those with no interest in repairing or improving their homes, adding to the splendour of their gardens, replacing that child and pet demolished sofa or attempting to snatch up the best bargains on the food shelves, there is always the option of shopping for that latest fashion.

Fashion in clothing and accessories changes so rapidly these days that many find it a struggle to stay abreast of what is *in* and what is *not*. It is considered by many that an item of clothing is required to sport one of the top logos in order to justify its purchase. Emblems ablaze on t-shirts, sport shoes and almost any form of apparel one can imagine. Hand bags, shoes, watches and trimmings must all display a name of fame for all to behold, and if it is also endorsed by a celebrity of note, then all the better.

Luckily many of us cannot afford such an indulgence so do not bother, others simply do not care if this or that design is favoured by a skinny model whose name is instantly forgotten and so do not bother. Fashion is purely one of the best retail marketing ploys ever invented. Our choice of fashion in finality depends on who we are trying to impress, not what we may choose for ourselves if the act of honesty were involved, and it is a fact that the predominance of our fashion preferences are made to gain the attraction of another. Let's be truthful here, the majority of the clothes wearing populous would not consider wearing a garment that caused that person of our hearts desire to collapse in fits of hysterical laughter!

Our attraction to another has long been the subject of discussion, contemplation and even ridicule. Who we as individuals choose as a life partner or spouse depends on many factors, not just a pretty face or a bulging wallet. We all experience desire for that beautiful film starlet or that handsome young male singer but these are aspiration, hopes that dreams are made of. However in reality we mostly combine with others of similar wealth, or lack of, good looks or lack of, personality and possibly even hereditary family status. There cannot be one lone reason for being attracted to another; rather it is a boiling pot of assorted ingredients that finally becomes an attraction between two persons.

Class or breeding can play a part in the attraction of one to another and is still a very important factor in many countries, princesses vie to marry princes, and aristocrats of noble birth seek to prolong their heritage by marrying into a family of similar wealth and prominence.

The definition of social class may be considered on the decline but in truth it is not, however it is possible that what was once considered as the class structure has now become diluted or possibly tainted by wealth gained through business, fame or chance. Not all countries support a monarchy with its royalty, lords, ladies and pouting princes alongside preening princesses but in those countries that still retain a King or Queen, class and breeding is still an important feature in those societies.

But who are *they*? Well we all know who others are referring to whenever the title of *they* crops up in a conversation. The majority of the world wide populous use this term in relation to those in authority, those in charge of some aspect of our lives and those who's professional titles are too long winded to be repeated more than once. No matter how infuriating the continued use of the term *they* is, it is now so commonly used that the word *they* has far outgrown its initial intention. *They* are watching us, *they* are in charge, *they* decide, *they* are the government, the medical profession, the councils, the lawyers and even the dustmen. *They* is the designation for all.

And now for some music. From the earliest of times when man first knocked two sticks together and discovered a sound. By striking the two sticks in a repeating order, they discovered a beat. By vocalising sounds along side the rhythmic beat, man discovered the first rudimentary music and probably frightened all nearby wild animals.

Today music is all around us and available in virtually any format imaginable. If it is electronic then there is a good chance it will play music; from stereo's to IPods, from personal computers to mp3 players, from phones to Walkman's, music blares into our lives constantly. Music is a main factor on television and radio with new singing stars constantly erupting like acne on a teenagers face, bursting forth upon our lives before flickering out again.

It is unusual to find some person who does not like any form of music, even the profoundly deaf enjoy music to a degree via the vibrations rippling on surfaces that can be touched by sensitive fingers. Admittedly there are some styles and formats of music which we may dislike, but none can honestly state that one does not appreciate any form of music.

We cannot define music itself but we can identify its properties. Music can make us laugh, it can make us cry, it can lift us up when we are down and music can calm the raging heart. Music will cause pleasant memories to resurface and tear at the heart strings of past loves. We cannot truly define music but possibly we can describe it, music is the food of the soul!

The impact of television upon our lives can be labelled as perhaps one of the most influential inventions of modern technology. Though we may decry our interest in television and deny that certain programmes entertain us, the vast majority will be transfixed to a television screen for a significant portion on our free time. Television influences many of us and we in turn, as viewers influence what subject matter the programmes consist of. There is no clear identification of the average viewer, rich or poor, academic or layman, famous or infamous; everyone is drawn towards the television at some point.

So there we sit, our gaze rooted upon the flickering silver screen positioned prominently within our homes, in many cases the television has spawned into other rooms throughout the house, and most bedrooms and kitchens now frequently boast the additional television. Children are lulled asleep by a familiar and friendly programme, and awake in the morning to a new day of television entertainment that will sustain their hopes, dreams and imagination throughout the day of boring reality. Television has enticed us all, be it through entertainment and information, amusement and controversy, music or news, all these avenues of interest spill forth from the brainchild of John Logie Baird.

Examining the trends and vices of modern living has not been easy; many facts and figures are already well established within the academic and medical societies. Categorizing our habits as vices may appear strong to many of those who partake in smoking, drinking, following fashion, over eating or doing drugs. How one concludes that a popular habit can be deemed a vice will largely depend on how we view our own existence and lifestyle. If one is of the opinion that their individual way of life is suitable then who are we to criticise? Tobacco, alcohol, excessive eating, following a fashion or using narcotics, each one of these vices is the choice of the individual, but sometimes the effects can spill into and change the lives of others.

In all walks of life and in all societies everyone seeks some form of release to their mundane lifestyle and we all seek out those routes of escape that will enable each one of us in our efforts to cope with, and understand the pressures of modern living. Similar to a pressure valve, we all need a diversion, a release or distraction from the worries, concerns, fears and strife that play a significant part in being human.

The modern human is a very strange beast indeed. Our priorities in life have changed as we strive to fulfil our own individual idealisms, hope and dreams. For centuries we humans have studied, dissected and analysed all aspects of human psychology, sociology and philosophy in great depth. Academics across the ages have speculated on behaviour, beliefs, fears and actions of the human animal and conclusions and assumptions have abounded throughout history. Great minds have sought the truth, the reasons for our existence and the real meaning of life for ions. Many have opened our minds to new possibilities, new visions and new insights into the profound depths of the human mind.

In the content of this book I have deliberately tried to avoid any subjects that require more than a simple superficial examination. Instead aspects and questions from today have been considered and discussed, a brief exploration of the modern Homo sapien, the actions and behaviour as witnessed by all but perhaps considered by none.

www.ingramcontent.com/pod-product-compliance
Lightning Source LLC
Chambersburg PA
CBHW061507180526
45171CB00001B/69